COMPUTATIONAL MODELS OF COGNITIVE PROCESSES

PROGRESS IN NEURAL PROCESSING*

Series Advisor
Alan Murray *(University of Edinburgh)*

Vol. 9:   RAM-Based Neural Networks
           Ed. James Austin

Vol. 10:   Neuromorphic Systems: Engineering Silicon from Neurobiology
           Eds. Leslie S. Smith & Alister Hamilton

Vol. 11:   Radial Basis Function Neural Networks with Sequential Learning
           Eds. N. Sundararajan, P. Saratchandran & Y.-W. Lu

Vol. 12:   Disorder Versus Order in Brain Function: Essays in Theoretical
           Neurobiology
           Eds. P. Århem, C. Blomberg & H. Liljenström

Vol. 13:   Business Applications of Neural Networks: The State-of-the-Art
           of Real-World Applications
           Eds. Paulo J. G. Lisboa, Bill Edisbury & Alfredo Vellido

Vol. 14:   Connectionist Models of Cognition and Perception
           Eds. John A. Bullinaria & Will Lowe

Vol. 15:   Connectionist Models of Cognition and Perception II
           Eds. Howard Bowman & Christophe Labiouse

Vol. 16:   Modeling Language, Cognition and Action
           Eds. Angelo Cangelosi, Guido Bugmann & Roman Borisyuk

Vol. 17:   From Associations to Rules: Connectionist Models of Behavior and
           Cognition
           Eds. Robert M. French & Elizabeth Thomas

Vol. 18:   Connectionist Models of Behaviour and Cognition II
           Eds. Julien Mayor, Nicolas Ruh & Kim Plunkett

Vol. 19:   Vision: Images, Signals and Neural Networks
           By Jenny Hérault

Vol. 20:   Connectionist Models of Neurocognition and Emergent Behavior
           Ed. Eddy J. Davelaar

Vol. 21:   Computational Models of Cognitive Processes
           Eds. Julien Mayor & Pablo Gomez

*To view the complete list of the published volumes in the series, please visit:
http://www.worldscientific.com/series/pnp

Proceedings of the 13th Neural Computation and Psychology Workshop

COMPUTATIONAL MODELS OF COGNITIVE PROCESSES

San Sebastian, Spain   12 – 14 July 2012

Editors

Julien Mayor
University of Geneva, Switzerland

Pablo Gomez
De Paul University, USA

World Scientific

NEW JERSEY · LONDON · SINGAPORE · BEIJING · SHANGHAI · HONG KONG · TAIPEI · CHENNAI

Published by

World Scientific Publishing Co. Pte. Ltd.

5 Toh Tuck Link, Singapore 596224

USA office: 27 Warren Street, Suite 401-402, Hackensack, NJ 07601

UK office: 57 Shelton Street, Covent Garden, London WC2H 9HE

Library of Congress Cataloging-in-Publication Data
Neural Computation and Psychology Workshop (13th : 2010 : London, England)
    Computational models of cognitive processes : proceedings of the 13th Neural Computation and
Psychology Workshop (NCPW13) / by Julien Mayor (University of Geneva, Switzerland) &
Pablo Gomez (De Paul University, USA).
    pages cm. -- (Progress in neural processing ; vol. 21)
    Includes bibliographical references and index.
    ISBN 978-9814458832 (hardcover : alk. paper)
    1. Neural networks (Neurobiology)--Congresses. 2. Cognition--Congresses. 3. Neural
stimulation--Congresses. I. Mayor, Julien, editor of compilation. II. Gomez, Pablo (Pablo Alegria),
editor of compilation. III. Title. IV. Title: NCPW13.
    QP363.3.N395 2010
    612.8'233--dc23
                                                        2013039707

British Library Cataloguing-in-Publication Data
A catalogue record for this book is available from the British Library.

In-house Editor: Sandhya Devi

Printed in Singapore

Preface

From July 12th to 14th 2012, the 13th Neural Computation and Psychology Workshop (NCPW13) took place in San Sebastian, Spain. Over three days, this well-established and lively workshop brought together researchers from different disciplines such as artificial intelligence, cognitive science, computer science, neurobiology, philosophy and psychology to discuss their work on models of cognitive processes.

In line with its tradition, the workshop was characterised by its limited size, high quality papers, the absence of parallel talk sessions, and a schedule explicitly designed to encourage interaction among the researchers present in an informal setting.

Based on this very successful event, this volume provides a snapshot of communications and posters that were presented during the conference. The diversity of topics in cognitive science discussed in the conference is mirrored by the diversity of articles featured in this volume: from language acquisition to problem solving to semantic cognition and reading.

We hope that this volume will distill the same enthusiasm that was present throughout the conference.

We would like here to take the opportunity to thank all attendees for making this conference a great success, with a special thanks to our invited speakers; professors Jay McClelland, Mark Seidenberg and Randall O'Reilly. We would also like to extend our gratitude to the generosity of Jay McClelland, who created and funded the Rumelhart Memorial Travel Awards, sponsoring nine very bright students from around the world so that they would be able to attend NCPW13. Congratulations to them!

Finally we would like to acknowledge the many forms of support that we benefited from during NCPW13:

The conference and this volume could not have seen light without the financial support of the EUCogIII society, of the Government of the Basque Country and of the University of the Basque Country, via its summer course program. JM would like to acknowledge his funding source; grant 131700 awarded by the Swiss National Science Foundation.

Furthermore, organising this conference would not have been possible without the expert help of the BCBL administration staff; in particular Leire Arietaleanizbeascoa, Pawel Kuszelewski and Jon Orozco. Eskerrik Asko! We would also like to thank warmly Manuel Carreiras, director of the BCBL, for hosting this conference and providing all the necessary support and resources for organising the 13th NCPW conference in the best possible conditions.

We already look forward to the next edition and wish the readers all the best in their journey through the different contributions to neural computations in psychology.

Julien Mayor and Pablo Gomez, NCPW13 organisers.

Contents

Preface   v

Language

Modelling Language – Vision Interactions in the Hub and
Spoke Framework
A. C. Smith, P. Monaghan and F. Huettig   3

Modelling Letter Perception: The Effect of Supervision and
Top-Down Information on Simulated Reaction Times
M. Klein, S. Frank, S. Madec and J. Grainger   17

Encoding Words into a Potts Attractor Network
S. Pirmoradian and A. Treves   29

Unexpected Predictability in the Hawaiian Passive
'Ō. Parker Jones and J. Mayor   43

Difference Between Spoken and Written Language Based on
Zipf's Law Analysis
J. S. Kim, C. Y. Lee and B. T. Zhang   62

Reading Aloud is Quicker than Reading Silently: A Study in the
Japanese Language Demonstrating the Enhancement of Cognitive
Processing by Action
H.-F. Yanai, T. Konno and A. Enjyoji   72

Development

Testing a Dynamic Neural Field Model of Children's
Category Labelling
K. E. Twomey and J. S. Horst   83

Theoretical and Computational Limitations in Simulating 3- to
4-Month-Old Infants' Categorization Processes
M. Mermillod, N. Vermeulen, G. Kaminsky,
E. Gentaz and P. Bonin   95

Reinforcement-Modulated Self-Organization in Infant Motor
Speech Learning
A. S. Warlaumont   113

A Computational Model of the Headturn Preference Procedure:
Design, Challenges, and Insights
C. Bergmann, L. Ten bosch and L. Boves   125

Right Otitis Media in Early Childhood and Language Development:
An ERP Study
M. F. Alonso, P. Uclés and P. Saz   137

High-Level Cognition

The Influence of Implementation on "Hub" Models of
Semantic Cognition
O. Guest, R. P. Cooper and E. J. Davelaar   155

Hierarchical Structure in Prefrontal Cortex Improves Performance
at Abstract Tasks
R. Tukker, A. C. Van Rossum, S. L. Frank and W. F. G. Haselager   170

Interactive Activation Networks for Modelling Problem Solving
P. Monaghan, T. Ormerod and U. N. Sio   185

On Observational Learning of Hierarchies in Sequential Tasks:
A Dynamic Neural Field Model
E. Sousa, W. Erlhagen and E. Bicho   196

Knowing When to Quit on Unlearnable Problems: Another
Step Towards Autonomous Learning
T. R. Shultz and E. Doty   211

A Conflict/Control-Loop Hypothesis of Hemispheric Brain
Reserve Capacity
N. Rendell and E. J. Davelaar   222

Action and Emotion

Modeling the Actor-Critic Architecture by Combining Recent
Work in Reservoir Computing and Temporal Difference Learning
in Complex Environments
J. J. Rodny and D. C. Noelle   237

The Conceptualisation of Emotion Qualia: Semantic Clustering
of Emotional Tweets
E. Y. Bann and J. J. Bryson   249

A Neuro-Computational Study of Laughter
M. F. Alonso, P. Loste, J. Navarro, R. Del Moral,
R. Lahoz-Beltra and P. C Marijuán   264

Language

MODELLING LANGUAGE – VISION INTERACTIONS IN THE HUB AND SPOKE FRAMEWORK

A. C. SMITH

Max Planck Institute for Psycholinguistics
Nijmegen, The Netherlands

P. MONAGHAN

Department of Psychology, Lancaster University
Lancaster LA1 4YF, UK

F. HUETTIG

Max Planck Institute for Psycholinguistics
Nijmegen, The Netherlands

Multimodal integration is a central characteristic of human cognition. However our understanding of the interaction between modalities and its influence on behaviour is still in its infancy. This paper examines the value of the Hub & Spoke framework[1-4] as a tool for exploring multimodal interaction in cognition. We present a Hub and Spoke model of language–vision information interaction and report the model's ability to replicate a range of phonological, visual and semantic similarity word-level effects reported in the Visual World Paradigm.[5,6] The model provides an explicit connection between the percepts of language and the distribution of eye gaze and demonstrates the scope of the Hub-and-Spoke architectural framework by modelling new aspects of multimodal cognition.

1. Introduction

Hub and Spoke (H&S) models[1-4] are characterised by a central resource that integrates modality specific information. The approach reflects the increased interest and awareness within cognitive science of multimodal cognitive interactions. To date computational implementations of the H&S framework have been used in conjunction with neuropsychological data to offer both explanation for a range of semantic related neuropsychological disorders and insight into how semantic processing may be implemented within the brain. This paper aims to highlight the potential for broader application of this framework. Research within cognitive neuroscience has demonstrated the difficulty of assigning modality specific functions to distinct neural processing regions (see Anderson[7]; Poldrack[8]). An increased understanding of how modality specific information may be integrated with information from other modalities and how such a system may behave could therefore prove valuable to neuropsychology. The H&S computational modelling framework offers a tool for investigating such complex interactive aspects of multimodal cognition.

2. Virtues of the Hub & Spoke Framework

When hearing a spoken word such as "apple" it is possible to bring to mind its visual form. When seeing an object such as an 'apple' it is also possible to bring to mind the spoken word used to describe the object "apple". How are modality specific representations connected across modalities, what is the nature of representation in each modality and how are connections between representations acquired? Previous H&S models have offered answers to each of these questions.

The H&S framework has proved successful by providing a parsimonious architecture in which single modality models can be drawn together to examine the consequences of multimodal interaction and representation. Due to the complexity inherent in multimodal processing, predicting the connectivity between modalities without explicit implementation can be challenging. For instance, an apparent dissociation between lexical and semantic performance in semantic dementia patients suggested the need for separate systems supporting lexical and semantic processing. However, the H&S offered a means of testing the compatibility of a fully integrated model with the behavioural data, and Dilkina et al.[3,4] demonstrated that counter to previous assumptions the pattern of behaviour observed was consistent with a single system H&S model.

The H&S framework offers a single system architecture with minimal architectural assumptions, and this makes it possible to isolate the influence of two further major determinants of emergent behaviour in such complex multimodal systems, 1) the structure of representations and/or 2) the tasks or mappings demanded by the learning environment.

Plaut[1] and Rogers et al[2] present two alternative means of exploring the role of representational structure through use of the H&S framework. Plaut[1] focused on a single aspect of representational structure, specifically systematic or arbitrary relationships between modalities. By abstracting out additional complexity within representations Plaut was able to investigate the emergent properties of this single factor. In contrast Rogers et al[2] and Dilkina et al[3,4] provided a richer representation of the structure available within the learning environment by deriving semantic representations from attribute norming studies. This enabled the authors to examine the emergent properties a single system multimodal model is capable of developing with such richer input. It is through simulating such complexity within the learning environment that their model was able to replicate the broad variability displayed by semantic dementia patients that had previously been viewed as challenging for single system accounts. Such approaches demonstrate the framework's potential for providing a more detailed understanding of how representational structure shapes multimodal cognition.

As the H&S framework allows a model to perform multiple mappings, decisions are required as to which mappings are performed, how frequently they are performed and how these variables might change over the course of development. Dilkina et al.[4] introduced stages of development within the model training process. They attempted to provide a more accurate depiction of the constraints placed on systems during development by manipulating the frequency and period in which given tasks are performed by the model (e.g., mapping orthography to phonology was only performed during the second stage of development). This is an example of how the framework can be used to explore the relationship between environmental constraints, such as the type and frequency of mappings performed during development, and the emergent behaviour displayed by the system.

To date the H&S framework has been used primarily in conjunction with neuropsychological data. This approach provides clear advantages when aiming to map network architecture onto neural populations and has brought significant progress in this direction with evidence emerging for a mapping of the semantic hub (integrative layer) onto neural populations in the anterior temporal lobe.[9] The framework however also offers scope for examining the factors underlying individual differences within non-patient populations, a feature yet to be exploited. For example, as we have described, the framework makes it possible to examine how contrasts in the learning environment, be it in the input to the system (e.g., richness or diversity of input) or the mappings demanded (e.g., learning to read: orthography to phonology), can result in variation in behaviour both across development and in mature systems.

Further as multimodal integration is central to many aspects of human cognition the H&S framework has the potential to provide insight into many new areas of cognitive processing. The following section provides an example, utilizing the framework to model the influence of language-vision interactions on eye gaze behaviour.

3. A Hub & Spoke Model of Language Mediated Visual Attention

3.1. *Language Mediated Visual Attention & The Visual World Paradigm*

Within daily communication we experience the cognitive system's ability to rapidly and automatically integrate information from linguistic and non-linguistic streams. For example many spoken utterances become ambiguous without the context in which they are delivered (e.g., "I love this field!"). Many have placed profound importance on the role of multimodal interaction during language processing[10], arguing that it is such multimodal context that grounds communication in the world that we share.

The Visual World Paradigm (VWP) offers a means of examining online language-vision interactions and has led to a substantial literature examining the role of non-linguistic information in language processing (for review see Huettig, Rommers, and Meyer[11]). Within the paradigm, participants are exposed to a visual display and an auditory stimulus while their eye movements are recorded. Its application has led to a detailed description of how manipulation of relationships between visual and auditory stimuli can drive systematic variation in eye gaze behaviour.

Allopenna et al.[12] demonstrated that the probability of fixating items within the visual display in a VWP could be modulated by phonological overlap between the name corresponding to the item depicted and the target word in the auditory stimulus. Given a target word (e.g. beaker), cohort competitors (e.g. beetle) were fixated earlier and with higher probability than rhyme competitors (e.g. speaker), both of which were fixated with higher probability than unrelated distractors (e.g. carriage). Simulations using the TRACE model of speech perception[13] replicated both the time course and probability of fixations displayed by participants.

Visual relationships between spoken words and displayed items have also been shown to influence fixation behaviour in the VWP. Dahan and Tanenhaus[14] and Huettig and Altmann[15] presented within a visual display items that shared visual features (e.g. snake) with target items (e.g. rope). They observed that participants fixated such visual competitors with higher probability than unrelated distractors both when the target was present in the display[14] and when the target was absent[15].

Semantics provides a third dimension in which overlap between target and competitor has been shown to modulate fixation behaviour. Huettig and Altmann[16] and Yee and Sedivy[17] both demonstrated that items within the visual display that shared overlapping sematic features with a target word were fixated with higher probability than unrelated distractors. Again this was demonstrated in conditions in which the target was present[17] and when the target was absent.[16]

Taken together this evidence indicates that language mediated eye gaze can be driven by representational overlap in visual, semantic and phonological dimensions. Although such studies provide a detailed description of how eye gaze varies as a function of the relationships between visual and auditory stimuli we know very little about the processes connecting these events. Issues such as what is the nature of representation in each modality, how are representations across modalities connected and how is activation of such representations connected to eye gaze behaviour remain unresolved.

The behaviour of participants in the VWP is dependent on language-vision interactions, the interaction of visual information from the visual display, and auditory information from a spoken utterance. Although models of language vision interaction applied to the VWP[10,18,19] exist, they lack sufficient depth of representation in visual, semantic and phonological dimensions to provide explanation for the range of word level effects observed in the VWP. Similarly single modality models of VWP[12,20] although providing depth of representation in a single modality, lack the necessary representation in other modalities that is required to offer a comprehensive description of language mediated eye gaze.

Our current study aims to explore the scope of the H&S framework as a tool for examining multimodal cognition by using it to derive a model of language mediated eye gaze that provides an explicit connection between the percepts of language and the distribution of eye gaze. We constructed a recurrent neural network that integrates semantic, visual and phonological information within a central resource and test its ability to capture the range of word level effects observed in the VWP, described in this section.

3.2. Method

3.2.1. Network

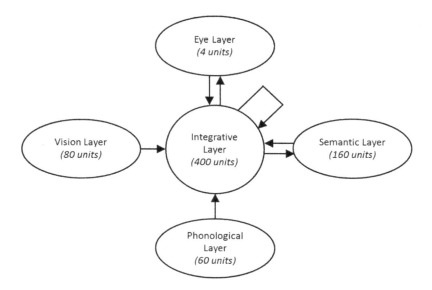

Figure 1. Network Architecture

The architecture of the neural network used in this study is displayed in Figure 1. It consists of a central resource that integrates modality specific information from four visible layers. The phonological layer is composed of six phoneme slots, each represented by 10 units and models the input of phonological information extracted from the speech signal. The vision layer encodes visual information extracted from four locations in the visual display each location represented by 20 units. The central integrative layer is self-connected and has both feed-forward and feed-back connections to both semantic and eye layers. The semantic layer consists of 160 units and allows the model to capture the semantic properties of items. The eye layer is defined by 4 units, with each unit associated with a given location in the visual field. Activation of each eye unit is taken to represent the probability of fixating the associated spatial location in the visual display. Layers are fully connected in all specified directions shown in Figure 1.

3.2.2. Artificial Corpus

The artificial corpus was constructed taking a fundamentalist approach (see Plaut[1]) to allow control over relationships between representations embedded within and across modalities. The corpus consisted of 200 items each with unique phonological, visual and semantic representations.

Phonological representations were produced by pseudo-randomly assigning a fixed sequence of six phonemes to each word with each phoneme taken from a phoneme inventory consisting of 20 possible phonemes. Each phoneme was defined by a 10 unit binary vector. As was the case with semantic and visual representations binary values were assigned with $p = 0.5$. Visual representations were defined by 20 unit binary vectors. Semantic representations were encoded by 160 unit binary vectors. Semantic representations were sparsely distributed, with each item assigned 8 of a possible 160 semantic properties.

Table 1. Controls on signal overlap within artificial corpus

Modality	Item	Constraint	Signal Overlap $(\bar{x})$
Phonology	Competitor	3 of 6 phonemes shared with target	75%
	Unrelated	Max. 2 consecutive phonemes shared with any other item	50%
Function	Near	4 of 8 functional properties shared with target	97.5%
	Unrelated	Max. 1 functional property shared with target	95%
Vision	Competitor	For 10 visual features P(feature overlap with target) = 1; for remaining features P(feature overlap with target) = 0.5	75%
	Unrelated	P(feature overlap with target) = 0.5	50%

Details of the constraints on the construction of representations applied within each modality can be found in Table 1. Onset competitors shared their initial sequence of three phonemes with target items, while rhyme competitors shared the final sequence of three phonemes with targets. All other phonological representations overlapped by a maximum of two consecutive phonemes. Visual competitors shared 10 visual features with targets and overlapped with $p = 0.5$ across the remaining 10 features. Semantic neighbours shared 4 of 8 semantic properties with target items.

3.2.3. Training

The model was trained on four tasks chosen to simulate the tasks performed by participants prior to testing through which associations between representations are acquired. The model was trained to map from visual representations to semantic representations, from phonological representations to semantic representations and to activate the eye unit corresponding to the location of the visual representation of a target defined by the presence of its phonological or semantic form. Details of the timing of events within each training task can be found in Table 2. In tasks that involved the presentation of phonological information phonemes were presented to the model sequentially, with one additional phoneme presented at each subsequent time point.

Training tasks were assigned trial by trial on a pseudo random basis. To reflect our assumption that individuals select items based on their associated semantic properties more frequently than selecting items based on an external auditory stimulus, phonology driven orienting tasks occurred four times less than all other training tasks. Items were selected from the corpus and assigned locations and/or roles (target/distractor) randomly. All connection weights were initially randomised and adjusted during training using recurrent back-propagation (learning rate = 0.05). Training was terminated after 1 million trials.

Table 2. Temporal organisation of events in model training

Task	Vision (Vis)		Phonology (Phon)		Semantics (Sem)		Eye	
	Description	ts	Description	ts	Description	ts	Description	ts
Vis to Sem	4 visual representations randomly selected from corpus, 1 assigned as target	0-14	Random time invariant noise provided as input	0-14	*Semantic representation of target provided post display onset*	*3-14*	Location of target activated, all other locations inactive	0-14
Phon to Sem	Random time invariant noise provided as input across all 4 input slots	0–14	Phonology of target provided as a staggered input	0-14	*Semantic representation of target provided post phonological disambiguation*	*5-14*	No constraints on activation	-
Phon to Location	4 visual representations randomly selected from corpus, 1 assigned as target	0-14	Phonology of target provided as a staggered input	0-14	No constraints on activation	-	*Post disambiguation location of target activated, all other locations inactive*	*5-14*
Sem to Location	4 visual representations randomly selected from corpus, 1 assigned as target	0-14	Random time invariant noise provided as input	0-14	Semantic representation of target provided	0-14	*Location of target activated, all other locations inactive post functional onset*	*2-14*

3.2.4. *Pre-Test*

Results presented in this paper reflect the mean performance of six instantiations of the model. Once trained the model was tested on its ability to perform each of the four training tasks for all items in all locations. In mapping from visual or phonological form to semantics, activation of the semantic layer was most similar (cosine similarity) to that of the target for 100% of items. For both phonological and semantic orientation tasks the eye unit associated with the location corresponding to that of the target was most active for 98% of items.

3.3. *Results*

Within the following simulations input to the visual layer was provided at time step (ts) 0 and remained until the end of each trial (ts 29). Figures display the average activation of the eye unit corresponding to the location of the given competitor type as a proportion of the total activation of all units within the eye layer. Onset of the target phonology occurs at ts 5, with a single additional phoneme provided at each subsequent ts. For each simulation 20 experimental sets were constructed, with each set containing 4 items. The model was then tested on all 20 sets with each set tested in all 24 possible arrangements of item and location. Comparisons of the probability of fixating each competitor type were calculated using the procedure described in Huettig & McQueen.[21] The mean probability of fixating a given type of competitor was calculated across all time points and across all trials in a given set. The ratio between the mean probability of fixating a given competitor type and the sum of the probability of fixating the given competitor and the competitor against which it is to be compared was calculated. This ratio was then compared across sets to 0.5 using a one-sample t-test.

3.3.1. *Simulation of Phonological Effects*

To simulate Allopenna et al[12] the model was presented with a display containing the visual representation of a target, an onset competitor (first three phonemes overlap with target), a rhyme competitor (final three phonemes overlap with target) and an unrelated distractor (no phonological overlap with target) (see Figure 2A). All items were controlled for semantic and visual similarity. Both onset competitors [mean ratio = 0.57, $t(19) = 9.86$, $p < 0.001$] and targets [mean ratio = 0.76, $t(19) = 54.84$, $p < 0.001$] were fixated with higher probability than unrelated distractors, although onset competitors were fixated less than the targets [mean ratio = 0.30, $t(19) = -56.91$, $p < 0.001$]. These patterns of fixation behaviour replicate those reported in Allopenna et al.[12] However in contrast to their findings the probability of fixating rhyme competitors did not differ from that of unrelated distractors.

Although such conditions were not tested in Allopenna et al[12] simulations were also run in which the target item was replaced by an additional unrelated distractor (Figure 2B). As in the target present condition onset competitors were fixated more than distractors [mean ratio = 0.57, $t_2(19)$ = 13.11, p < 0.001], while the probability of fixating rhyme competitors and unrelated distractors did not differ.

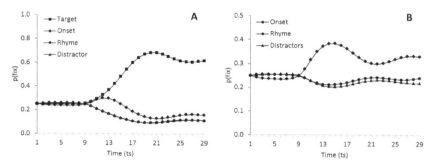

Figure 2. Mean proportional activation of eye units across all items and locations A) phonological competitors & target item; B) phonological competitors only (target absent)

3.3.2. *Simulation of Visual Effects*

In Dahan & Tanenhaus[14] participants are presented with displays containing a target, a visual competitor and two unrelated distractors, these conditions were simulated within the model with results presented in Figure 3A. The model replicates the behaviour of participants reported in Dahan & Tanenhaus[14] with both target items [mean ratio = 0.76, t(19) = 82.70, p <0.001] and visual competitors [mean ratio = 0.61, t(19) = 23.97, p <0.001] fixated with higher probability than unrelated distractors. The model also fixated visual competitors [mean ratio = 0.33, t(19) = -47.19, p <0.001] less than target items again replicating the behaviour of participants.

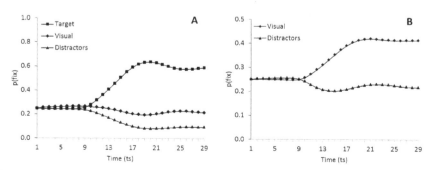

Figure 3. Mean proportional activation of eye units across all items and locations A) visual competitor & target item; B) visual competitor only (target absent)

Huettig and Altmann[15] tested participants using displays containing a single visual competitor in addition to three unrelated distractors, this was also simulated in the model with results shown in Figure 3B. Again the model successfully replicates the behaviour displayed by participants, fixating visual competitors with higher probability than unrelated items [mean ratio = 0.60, t(19) = 24.43, p <0.001].

3.3.3. Simulation of Semantic Effects

Simulations were also run of conditions reported in Yee and Sedivy[17] and Huettig and Altmann[16] in which participants are presented with scenes containing a target, a semantic competitor and two unrelated distractors (see Figure 4A). The model replicated participants increased probability of fixating target items [mean ratio = 0.76, t(19) = 96.44, p <0.001] and semantic competitors [mean ratio = 0.58, t(19) = 12.73, p <0.001] in comparison to unrelated distractors. Also as with participants semantic competitors were fixated less that target items [mean ratio = 0.30, t(19) = -50.04, p <0.001].

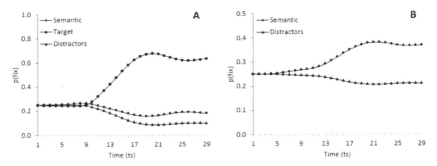

Figure 4. Mean proportional activation of eye units across all items and locations A) semantic competitor & target item; B) semantic competitor only (target absent)

Huettig & Altmann[16] also report the participants' fixation behaviour when presented with scenes in which the target is replaced by an additional unrelated distractor. Again the model replicates the pattern of increased fixation towards sematic competitors over unrelated items [mean ratio = 0.58, t(19) = 14.46, p <0.001] reported by Huettig and Altmann.[16]

4. Discussion

The results demonstrated that the model of language – vision interactions proposed within this paper is capable of replicating a broad range of single modality, word-level effects reported within the VWP literature. The model successfully replicates phonological onset, visual and semantic competitor effects in both target present and target absent conditions.

Although it does not precisely simulate the rhyme effects observed in Allopenna et al[12], we do not believe that this constitutes a significant problem for the proposed framework. Rhyme effects reported within the literature tend to be weak and far less robust than the onset effects the model does successfully replicate (see Allopenna et al[12]; Desroches et al[22]; McQueen & Viebahn[23]; McQueen & Huettig[24]). Moreover McQueen & Huettig[24] provides evidence that the comparative onset rhyme effect is modulated by the level of intermittent noise present in the speech signal. It is therefore possible such subtle effects are beyond the scope of the current model. However we do offer an alternative explanation in which the current framework remains compatible with the observed rhyme effects. In the current model input is a perfect replication of the corresponding auditory or visual representation. Therefore, initial phonological input always corresponds to the target item. However for real world stimuli the perceptual system is frequently provided with impoverished or noisy representations. We believe that training the model on a more realistic representation of the stimuli available in the true learning environment may lead to the emergence of rhyme effects. A model trained on noisy auditory input would rely less wholly on early sections of the speech signal, for sections of this signal may not correspond to the target representation. Consequently in comparison to the current model, greater value would be placed on later sections of the signal as they may contain additional valuable information that would allow the model to identify the target. Future work could explore this aspect of the learning environment through the addition of noise to model input; we predict that its introduction would reduce the influence on attention of initial phonemes and make it more likely for attention to be drawn by items with consistent overlap in later phonemes.

The above study provides an example of how single modality models can be integrated through the H&S framework to explore multimodal cognitive interactions. Although the connectivity between modalities is likely to be far more complex than that captured by the current model[25], our results indicate that the current framework can act as a good initial proxy for this interaction.

Given the models success in replicating the VWP data this would suggest that the model successfully captures some true aspect of the underlying cognitive processes recruited during completion of such tasks. The model presented offers an explicit description of the connection between the percepts of language and the distribution of eye gaze. This includes a description of the structure of representation within each modality, the features of those representations that drive competitor effects, the mechanisms that connect representations across modalities and a description of how such connections emerge from the structure of the learning environment. VWP research has moved on to examining multimodal effects in language-vision interactions (see Huettig & McQueen[21]). We hope to extend the current study by exploring the extent to which hypotheses implemented within the above model also offer explanation for such multimodal effects.

A further stage in this research project will be to exploit additional virtues of the H&S framework described in earlier sections of this paper to investigate the mechanisms that underlie individual differences in language mediated eye gaze. Recent evidence suggests that environmental factors such as exposure to formal literacy training can have a significant effect on language mediated visual attention.[26] Through use of the current model we aim to examine which environmental factors could give rise to the variation in behaviour observed and provide an explicit description of how such factors drive variation in eye gaze behaviour.

References

1. D. C. Plaut, *Cognitive Neuropsychology* **19**, 603 (2002).
2. T. T. Rogers, M. A. Lambon Ralph, P. Garrard, S. Bozeat, J. L. McClelland, J. R. Hodges, and K. Patterson, *Psychological Review* **111**, 205 (2004)
3. K. Dilkina, J. L. McClelland, D. C. Plaut, *Cognitive Neuropsychology* **25**, 136 (2008).
4. K. Dilkina,J. L. McClelland and D. C. Plaut, *Brain Research* **1365**, 66 (2010)
5. R. M. Cooper, *Cognitive Psychology* **6**, 84 (1974).
6. M. K. Tanenhaus, M. J. Spivey-Knowlton, K. M. Eberhard and J. C> Sedivy, *Science* **268**, 1632 (1995).
7. M. L. Anderson, *Behavioural and Brain Sciences* **33**, 245 (2010).
8. R. A. Poldrack, *Trends in Cognitive Sciences* **10**, 59 (2006).
9. M. A. Lambon-Ralph, K. Sage, R. W. Jones and E. J. Mayberry, E. J., *PNAS* **107**, 2717 (2010).
10. M. R. Mayberry, M. W. Crocker and P. Knoeferle, *Cognitive science* **33**, 449 (2009).
11. F. Huettig, J. Rommers and A. Meyer, *Acta Psychologica* **137**, 151 (2011).
12. P. D. Allopenna, J. S. Magnuson and M. K. Tanenhaus, M., K., *Journal of Memory and Language* **38**, 419 (1998).
13. J. L. McClelland and J. L. Elman, *Cognitive Psychology* **18**, 1 (1986).

14. D. Dahan and M. K. Tanenhaus, *Psychonomic Bulletin & Review* **12**, 453 (2005).
15. F. Huettig and G. T. M. Altmann, *Visual Cognition* **15**, 985 (2007).
16. F. Huettig and G. T. M. Altmann, *Cognition* **96**, B23 (2005).
17. E. Yee and J. C. Sedivy, *Journal of Experimental Psychology: Learning, Memory, and Cognition* **32**, 1 (2006).
18. M. Spivey, *The continuity of mind* (Vol. 40). Oxford University Press, USA (2008).
19. A. Kukona and W. Tabor, *Cognitive Science* **35**, 1009 (2011).
20. D. Mirman and J. S. Magnuson, *Memory & cognition* **37**, 1026 (2009).
21. F. Huettig and J. M. McQueen, *Journal of Memory and Language* **57**, 460 (2007).
22. A. S. Desroches, M. F. Joanisse and E. K. Robertson, *Cognition* **100**, B32 (2006).
23. J. M. McQueen and M. Viebahn, *Quarterly Journal of Experimental Psychology* **60**, 661 (2007).
24. J. M. McQueen and F. Huettig, *Journal of the Acoustical Society of America* **131**, 509 (2012).
25. C. McNorgan, J. Reid and K. McRae, *Cognition* **118**, 211 (2011).
26. F. Huettig, N. Singh and R. K. Mishra, *Frontiers in Psychology* **2**, 285 (2011).

MODELLING LETTER PERCEPTION: THE EFFECT OF SUPERVISION AND TOP-DOWN INFORMATION ON SIMULATED REACTION TIMES*

MICHAEL KLEIN

Laboratoire de Psychologie Cognitive, CNRS & Aix-Marseille University
3, place Victor Hugo, Bat. 9, Case D, 13331 Marseille
E-mail: Michael.Q.Klein@gmail.com

STEFAN FRANK

Centre for Language Studies, Radboud University Nijmegen
Erasmusplein 1, 6525 HT Nijmegen
E-mail: s.frank@let.ru.nl

SYLVAIN MADEC AND JONATHAN GRAINGER

Laboratoire de Psychologie Cognitive, CNRS & Aix-Marseille University
3, place Victor Hugo, Bat. 9, Case D, 13331 Marseille
E-mail: {I.Jonathan.Grainger, Sylvain.Madec}@gmail.com

In this study, we model human letter-recognition times using neural networks that extract visual features from real images of the letters. We focus on learning, and on how different learning methods and other factors affect the correlation between simulated reaction times and behavioural data. Specifically, we are interested in studying the effect of 3 factors on this correlation: (i) utilisation of an error signal during learning (supervised vs. unsupervised learning), (ii) whether or not the letter labels exert a top-down influence on the extracted features, and (iii) the effect of letter frequencies. To do so, we used Restricted Boltzmann Machines (RBMs), Back-propagation networks, and RBM/Perceptron hybrid architectures. We find the highest correlations ($r = 0.67$) with supervised models when using top-down information of letter labels on the feature layer during training, but only when the letters' frequencies are taken into account during learning. This study shows that to account for human letter identification times, letter frequency seems to be the most important factor. In addition, top down information of letter labels on the extracted visual features appears to be essential (making the difference between a significant and non-significant correlation). Whether or not the model is supervised makes little difference in the correlation to human reaction time data, but fully unsuper-

*This work was funded by the ERC grant 230313.

vised models have more difficulty generating accurate categorisation for letters with very low frequencies.

Keywords: Restricted Boltzmann Machines; Back-Propagation; Letter recognition; Reaction Time Modelling

1. Introduction

In order not to be overwhelmed by massive unorganised sensory input, but to perceive the world in a meaningful way, humans categorise objects, events, and actions. Categorisation is beneficial, because unknown individual objects that fall into a known category tend to be of similar significance. To categorise an object a perceiver needs to understand which features of an object are relevant for it to be in a particular category and which features form irrelevant (e.g., random) variability. Recognising letters of the alphabet shares this general problem of similarity in variability. However, being two-dimensional and monochrome, letters present a tractable problem and are, thus, suitable material to study the cognitive processes involved in visual categorisation. Despite being among the more simple of categorisation problems, human letter recognition is far from understood. While there is a sheer intractable amount of experimental studies on letter perception dating back more than a hundred years (see Mueller & Weidemann (2012)[1] for a review), the number of explanatory computational models is still quite limited. While recognising isolated printed letters is not a hard problem for pattern recognition algorithms, there are few computational models that connect and explain human behavioural data. Possibly the best current model of human single letter perception[2] correlates simulated letter-perception times with significant peaks in the EEG signal. This model, however, does not learn and relies on a set of input features that are defined by the modeller.

Since learning and feature extraction are among the main strengths of neural models, it should be feasible to find a neural algorithm for learning letter recognition and feature extraction. However, it is far less simple to find a learning algorithm and architecture that are cognitively plausible and to build a model that can simulate human behavioural data. With letter recognition being a cortical process, and cortical learning being of the unsupervised Hebbian-type,[3,4] a good starting point appears to be unsupervised correlation learning. A Restricted Boltzmann Machine (RBM) uses unsupervised correlation learning and is also a good algorithm for feature extraction.[5]

In the study reported here, we use the RBM algorithm to extract letter features from images of letters, presented to the model in the form of binary pixels. This is combined with supervised learning into several architectures in order to vary two factors: (i) the top-down influence of the letter labels on the feature extraction and (ii) the location and time at which supervision is employed.

All these models are used to simulate reaction times by turning the output units into spiking neurons and counting the time steps until the first output unit spikes. The generated reaction times are correlated with the human reaction times reported in Madec et al. (2012).[6]

Furthermore, it has been reported that letter naming times correlate strongly with letter frequencies.[7] While this finding provides very important clues and constraints on the underlying representations and processes involved in letter perception, it is in itself not yet a (causal mechanistic) explanation. It is necessary to develop a cognitive model that can link relative exposure to letters during training to observable reaction times in letter recognition. To establish such a link, in this study all architectures are trained both with a training data set in which the number of items for each letter correspond with known French letter frequencies and with a training data set in which all letters are presented an equal number of times.

2. Method

2.1. *Simulations*

In this study, we tested four main architectures: (i) an RBM / perceptron hybrid, (ii) a pure RBM, (iii) an RBM fine-tuned with back-propagation, and (iv) a pure back-propagation model. The hybrid model (called *hybrid* because it combines two modules that use two different learning algorithms) uses the RBM algorithm to extract a layer of letter feature and then classifies those into letters using the supervised delta-rule (see figure 1A). This means that there is no effect of the letter labels on the features. The second architecture is a pure RBM, in which images and labels are presented on the input layer and a common hidden layer is trained (see figure 1B). For testing, only the image is presented and the hidden activation is computed. From the hidden activation the activation of the letter neurons is generated. In this architecture learning is fully unsupervised, but the labels have an influence on the emerging features on the hidden layer. The third architecture is identical to the second in the first half of the training.

Then, the back-propagation algorithm is used to fine-tune the weights (see figure 1C). In the final model, only back-propagation is used. All models are built, trained and tested with one, two, and three hidden layers. Every architecture is trained with letter frequencies, with logarithmic frequencies, and without any frequencies. This results in a $4 \times 3 \times 3$ experimental design. Fifty simulations (i.e., networks trained from scratch) were performed per cell of the design. We computed the correlations of the reaction times of every simulation with the reaction times of every other simulation within one cell. Henceforth, the average of all correlations within one cell is called the *internal correlation*.

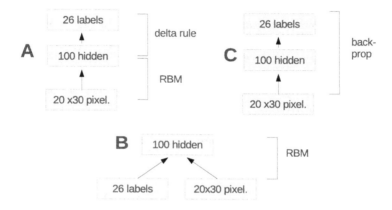

Fig. 1. Three different architectures tested in this study: (A) a hybrid model using unsupervised feature extraction with an RBM and then delta-rule training of the final classification layer; (B) a fully unsupervised RBM network with the labels having a top-down influence on the hidden layer; (C) a standard back-propagation network.

2.2. Neural Network Algorithms

2.2.1. Restricted Boltzmann Machines

An RBM[5] consists of an input and hidden layer, where every unit in the input layer is connected to every unit in the hidden layer, and each connection has symmetrical weights (i.e., the same value is used for bottom-up recognition and top-down down generation). There are no lateral connections between the units of a layer. In addition, every unit has a bias.

Unit activations are binary and stochastic: The activation of a unit j is set by first computing the sum of its weighted input, including its bias b_j:

$$z_j = b_j + \sum x_i w_{ij} \tag{1}$$

where x_i is the output of unit i and w_{ij} is the weight between units i and j.

From this sum, the probability that unit j receives an activation of 1 is computed using the logistic function:

$$P(x_j = 1|z_j) = \frac{1}{1 + e^{-z_j}} \tag{2}$$

For training, after random initialisation of weights and biases (here, using Gaussian distribution with mean 0 and standard deviation 0.1), an input vector is applied to the input layer and the hidden activations are computed from it. Then the input is reconstructed from the hidden representations by computing the downward activations. After that, a reconstructed hidden activation is computed from the reconstructed input (see Fig. 2).

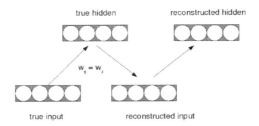

Fig. 2. The up and down algorithm: true hidden representations are generated from the true input. From the true hidden representations the input is reconstructed and from the reconstructed input the hidden representations are reconstructed.

Finally, the network learns by increasing the weights by the product of the input and the hidden units' activation minus the product of reconstructed input and hidden units:

$$\Delta w_{ij} = \epsilon((v_i h_j)_{\text{data}} - (v_i h_j)_{\text{rcon}}) \tag{3}$$

where $(v_i h_j)_{\text{data}}$ denotes the product of the input data v_i and resulting hidden activation h_j, and $(v_i h_j)_{\text{rcon}}$ is the same product but using the reconstructed activations. Both weights and biases change during learning. The weights of the input biases and hidden biases are changed accordingly.

The learning rule is analogue since biases can be treated as weights of connections from units that are always 1.

2.2.2. *Training a Deep-Belief Network*

Building up a fully unsupervised deep-belief network[8] is done in several steps. First, an RBM is trained on the input data with the algorithm described above. Next, the weights of this network are fixed. The input is then applied to this network with fixed weights and the output (at the hidden layer) serves as the input to the next RBM that is build on top of this one. This procedure continues until the desired depth of the overall neural model. Note, however, that to train a fully unsupervised deep-belief network, the final RBM does not only have the output of the pre-final RBM as input, but also the labels. To test such a model after training, the weights of the complete model have to be fixed. Then the input (without labels) is presented to the network. Finally the activation of the labels are generated top-down from the top layer of the model.

Subsequently, such a network can be fine-tuned with the back-propagation algorithm (explained below). To do so, however, the top-layer of the deep-belief network will serve as the pre-final layer and the representations of the labels will serve as a top-layer

2.2.3. *Delta-Rule and Back-Propagation*

The final layer of the hybrid model is trained using perceptron learning (delta rule). To fine-tune the weights of our deep-belief networks, we use the back-propagation algorithm.[9] A back-propagation network can be regarded as a multi-layer perceptron, or the perceptron as a special case (single-layer) of a back-propagation network. Therefore, we will treat both algorithms in this section.

The activation of a unit in both perceptron and back-propagation network is computed in the same manner as in the RBM (see equation 1). Also, the output y_j of a unit j is computed with the logistic function analog to the computation of the probability of a unit being 1 in the RBM. However, since the output is not binary, no sampling is necessary:

$$y_j = \frac{1}{1 + e^{-z_j}} \tag{4}$$

The computation of the error and the weight change for a perceptron is the same as for the final layer of a back-propagation network. In this kind of supervised learning, the output of the network y_k is subtracted from the desired output y_k^* to compute the error e_k. The weight change Δw_{ik} is then calculated by multiplying e_k by the activation x_i of the input neuron and the rate of change α:

$$e_k = y_k^* - y_k \tag{5}$$

$$\Delta w_{ik} = \alpha e_k x_i \tag{6}$$

For the back-propagation network the training of the weights to the pre-final layers is somewhat more complicated. While the rule for the update of the weights can stay the same as in the perceptron and the final layer, the computation of the weight change (Δw_{ik}) is quite different. We can only directly measure errors at the output layer. To get an approximation of the error at the hidden layer, we distribute the error of an output unit j to all the hidden units. The error is distributed proportional to the weight of the connection from the hidden unit to the output unit j. This assumes that the contribution of this hidden unit to the error is proportional to its connection strength to that output unit. The same algorithm applies to all the hidden layers of the network. Equation 7 shows the computation of the error at the hidden layer.

$$e_j = y_j(1 - y_j) \sum_k e_k w_{jk} \tag{7}$$

The error e_j at unit j is the sum of errors e_k of units k to which this hidden unit is projecting, times the weight w_{jk} of the connections multiplied with the output of the unit y_j times $(1 - y_j)$.

2.2.4. Simulating Reaction Times

To simulate reaction times after training the networks, we converted the neurons representing the letters into spiking leaky integrator neurons.[10,11] Every neural unit i was represented as a membrane potential u_i dynamically changing over time using the following equation:

$$u_i(t + 1) = u_i(t) + c \sum_k (w_{jk} y_k) - l; \tag{8}$$

The resting (and initial) value of the membrane potential was set to -70 (mV) for all units. Leakage l was set to 0.05 and a constant c used to convert input to the unit into voltage was set to 0.2. Reaction times were computed simply by counting the time steps until u_i reached the threshold of $-55mV$.

2.3. *Human Reaction Time Data*

The human reaction time data were taken from the study of Madec et al. (2012).[6] In this study, a behavioural index of letter identification processes was obtained by combining an immediate naming and a conditional delayed naming task (see Figure 3). In the immediate naming task, participants simply named as quickly as possible a target letter that was displayed on a computer screen. Then, on each trial, after naming the letter, participants performed a conditional delayed naming task. After a variable delay following their naming response, either a green or a red circle (light grey and dark grey in Fig. 3) was presented, and participants had to repeat the target letter's name they had just produced, only when they saw a green circle.

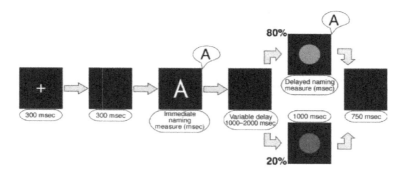

Fig. 3. Schematic depiction of the task in the Madec et al. (2012) study.

The immediate naming measure is assumed to include two main sources of variance that are related to the two main processes involved in letter naming. The first source of variance comes from visual identification processes, and the second source is related to output articulatory processes.

Having the measures of immediate naming and delayed naming, it is possible to compute a simple linear regression with immediate naming times being the dependent variable and delayed naming time being the independent variable (i.e., naming times are explained by delayed naming times). The residual values of the regression (i.e., the remaining unexplained variance), is likely to correspond to the time required by the visual identification processes.

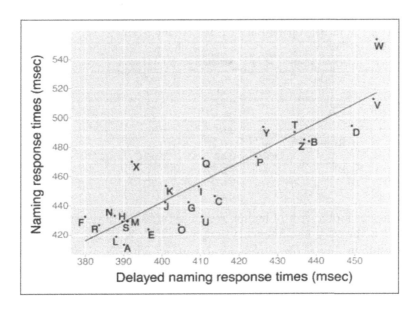

Fig. 4.   Immediate naming response times (msec) as a function of delayed naming response times (msec), reproduced from Madec et al. (2012).

3. Results

In all the following simulations, having more than one hidden layer yielded worse correlations with the reaction time data, rather than improving them. Hence, we will not give further details of the simulations with more than one hidden layer, but focus on the results of the architectures with one hidden layer.

Table 1 summarises the results: For each model, it shows the correlation between the simulated response times (averaged over the 50 runs) and human data (averaged over subjects), as well as the internal correlation (the correlation between different runs within the same architecture). Given the degrees of freedom, the correlation with response times is significant ($p < 0.05$) if $r > 0.38$, so three of the models simulate response times that significantly (and positively) correlate with the human response data. Importantly, these models also show very strong internal correlation, indicating that the results are reliable and do not strongly depend on the random initial connection weights or random sampling used in the RBM.

Significant correlations were only obtained when using a realistic letter-frequency distribution during training. This reflects the importance of let-

Table 1. Correlation of Simulated Reaction Times with Behavioural Data and Models' Internal Correlation

architecture	freq	correlation	int. correl.
hybrid	no	−0.22	0.09
	yes	0.21	0.09
RBM	no	0.13	0.23
	yes	0.61	0.8
	log	0.53	0.9
RBM + BP	no	−0.27	0.22
	yes	0.67	0.99
BP	no	−0.17	0.57
	yes	0.67	0.97

ter frequency as a predictor of human response times. However, in the pure RBM model, the very low frequency items (such as K, W, X, Y, and Z) did not result in any above-threshold activation. That is, they were not properly learned. We attempted to solve this problem by using logarithmically transformed letter frequencies (increasing the relative frequency of the infrequent letters). Although this was successful, it also reduced the correlation with human data to $r = 0.53$.

Another solution to the problem of low-frequency letter recognition turned out to be the fine-tuning of the RBM weights using back-propagation training. In addition, this increased the correlation with human response times to $r = 0.67$. For this model, the relation between simulated and actual response times (for each letter) is plotted in Figure 5.

When the model was trained with back-propagation from the beginning, the correlation was the same as in case of mixed RBM and back-propagation training ($r = 0.67$). Also, this model did respond to low-frequency letters. Hence, there was no benefit of applying the RBM prior to back-propagation.

The correlation between these latter models and the behavioural data ($r = 0.67$) were substantially higher than the correlation between (French) letter frequencies[7] and the behavioural data ($r = 0.49$), indicating that the model explains more than just the effect of letter frequencies.

4. Conclusions

In the study reported in this paper, we investigated the possibilities of modelling human letter perception with Restricted Boltzmann Machines. We found that fully unsupervised RBMs have problems learning low-frequency letters, but when trained with the logarithmic frequencies, they perform adequately and can be considered a valuable model of human letter per-

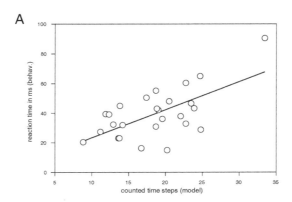

A

Fig. 5. Relation between simulated and behavioural data for RBM with back-prop fine-tuning.

ception. However, RBMs fine-tuned with back-propagation are superior to pure RBMs and, since pure back-propagation networks perform on the same level, it seems that nothing is gained by prior training with the RBM algorithm. Altogether, it appears that RBMs, while faster to train, are not superior to back-propagation networks when it comes to the modeling of human reaction time data. Furthermore, it seems that top down information of letter labels on the extracted visual features appear to be essential, since only the hybrid failed at simulating reaction times that showed significant correlations with human data. Finally, this study could confirm the essential role of letter frequency as a strong factor of letter recognition times and successfully distinguished cognitive architectures that link letter frequencies with reaction times from those that do not.

References

1. S. T. Mueller and C. T. Weidemann, *Acta Psychologica* **139**, 19 (2012).
2. A. Rey, S. Dufau, S. Massol and J. Grainger, *Cognitive Neuropsychology* **26**, 7 (2009).
3. R. Malinow, *Science* **252**, 722 (1991).
4. H. Markram, J. Lüubke, M. Frotscher and B. Sakmann, *Science* **275**, 213 (1997).
5. P. Smolensky, Information processing in dynamical systems: Foundations of harmony theory, in *Parallel Distributed Processing: Volume 1: Foundations*, eds. D. E. Rumelhart and J. L. McClelland (MIT Press, Cambridge, MA, 1986) pp. 194–281.

6. S. Madec, A. Rey, S. Dufau, M. Klein and J. Grainger, *J Cogn Neurosci* **24**, 1645 (2012).
7. B. New and J. Grainger, *Acta Psychologica* **138**, 322 (2011).
8. G. E. Hinton, *Trends in Cognitive Sciences* **11**, 428 (2007).
9. D. E. Rumelhart, G. E. Hinton and R. J. Williams, *Nature* **323**, 533 (1986).
10. L. Lapicque, *J Physiol Pathol Gen* **9**, 620 (1997).
11. A. N. Burkitt, *Biological Cybernetics* **95**, 1 (2006).

ENCODING WORDS INTO A POTTS ATTRACTOR NETWORK

SAHAR PIRMORADIAN, ALESSANDRO TREVES

SISSA, Cognitive Neuroscience Sector, Trieste, Italy

To understand the brain mechanisms underlying language phenomena, and sentence construction in particular, a number of approaches have been followed that are based on artificial neural networks, where words are encoded as distributed patterns of activity. Still, issues like the distinct encoding of semantic vs syntactic features, word binding, and the learning processes through which words come to be encoded that way, have remained tough challenges. We explore a novel approach to address these challenges, which focuses first on encoding words of an artificial language of intermediate complexity (BLISS) into a Potts attractor net. Such a network has the capability to spontaneously latch between attractor states, offering a simplified cortical model of sentence production. The network stores the BLISS vocabulary, and hopefully its grammar, in its semantic and syntactic subnetworks. Function and content words are encoded differently on the two subnetworks, as suggested by neuropsychological findings. We propose that a next step might describe the self-organization of a comparable representation of words through a model of a learning process.

Keywords: Word Representation, Artificial Language, Potts Attractor Network

1. Introduction

To understand the brain mechanisms underlying the phenomenon of language, specifically sentence construction, many studies have been done to implement an artificial neural network that encodes words and constructs sentences (see e.g. [1-4]). These attempts differ on how the sentence constituents (parts) are represented—either locally [1,3], or in a distributed fashion [5,6]—and on how these constituents are bound together—through either temporal synchrony [7], active circuits [3], or algebraic operations [8].

The local representation of each sentence constituent (either a word, a phrase, or even a proposition) results in an exponential growth in the number of units needed for structure representation [1]; this challenge was addressed in [3] by designing dynamic circuits between word assemblies, yet with a highly complex and meticulously (unrealistic) organized connections. In a fully distributed representation of words as vectors [5,6], words are bound (and *merged*) together by an algebraic operation—e.g. tensor product [8] or circular convolution [6]. Some steps have been attempted towards the neural implementation of such operations [4]. Another distributed approach was towards implementing a simple recurrent neural network that predicts the next word in a sentence [9]. Apart from the limited language size that the network could deal with [10], this system lacked an explicit representation of syntactic constituents, and it is shown it leads to a lack of grammatical knowledge in the network [11,3].

However, despite all these attempts, there remains the lack of a neural model that addresses the challenges of language size, semantic and syntactic distinction, word binding, and word implementation in a neurally plausible manner. We are exploring a novel approach to address these challenges, that involves encoding words of an artificial language of intermediate complexity into a neural network, as a simplified cortical model of sentence production, which stores the vocabulary and the grammar of the artificial language in a neurally plausible manner on two components: one semantic and one syntactic.

2. BLISS: The Training Language

As the training language of the network, we have constructed *BLISS* [12], for Basic Language Incorporating Syntax and Semantics. BLISS is a scaled-down synthetic language of intermediate complexity, which mimics natural languages by having a vocabulary, syntax, and semantics. Importantly, the degree of complexity of the language is designed having the size limitations of synthetic agents in mind, so as to allow for the use of equivalent corpora with human subjects and with computers, while aiming for reasonable linguistic plausibility.

BLISS is generated by a context-free grammar of limited complexity with about 40 production rules, with probabilities that were drawn from the *Wall Street Journal* (WSJ) corpus. It contains function words, inflectional suffixes, and some embedding structure. These grammatical features were introduced to enable experiments to investigate the ability for abstract pattern acquisition [13,14], the special role of function words [15], the role of suffixes [16], and especially hierarchical structures [17,18] in humans.

The BLISS vocabulary contains about 150 words, which belong to different lexical categories such as noun, verb, adjective, etc., and which were selected from the *Shakespeare* corpus. There are several studies investigating category learning in humans [19], and BLISS is intended to facilitate e.g. the analysis of the representation of distinct lexical categories.

Semantics is defined in BLISS as the statistical dependence of each word on other words in the same sentence, as determined solely by imposing constraints on word choice during sentence generation. We have applied different methods of weighing the preceding words so to determine which words come next. This should allow using BLISS to study at least rudimentary aspects of the emergence of semantic categories.

3. Potts Attractor Network: a Simplified Model of the Cortex

We have attempted to implement a neural network which mimics the neural mechanisms underlying sentence production. We use a *Potts associative memory network*, a generalization of an auto-associative memory network, an *attractor network* [20,21].

An attractor network is a collection of binary units that stores a concept—a pattern—in a distributed fashion, remembers a concept by completing a portion of it given as a cue, and uses a Hebbian learning rule to store a concept as an attractor at a minimum of the (free) energy of the network.

In the Potts associative memory network—the network of our interest—the units are not binary; instead, each can be activated in S different states. The Potts network has been proposed as a simplified model of macroscopic cortical dynamics [22,23], perhaps appropriate for modelling the language faculty and other high-level cognitive functions (Fig. 1(a)). The Potts network is a simplified two-level, local and global associative memory network [24,25], where a local network represents a patch of cortex, which locally stores features, and the global network associates those features to store concepts (as proposed by [26]). In the Potts network, the local associative memory networks are not described explicitly; instead, each is encapsulated each as a *Potts* unit. Thus, a Potts unit hypothetically models a patch of cortex, and the internal neuronal dynamics of the patch is not described by the model, rather it is subsumed into an effective description in terms of graded Potts units with adaptation effects. A collection of Potts units, connected through long-range synaptic connections, compose the global associative memory, which stores the concepts.

Apart from the simplification the Potts network offers, the choice of this network for sentence production has been mainly motivated by its "latching" dynamics [27]. Latching is an ability to jump spontaneously and in some conditions indefinitely from an attractor state to the next, in a process that mimics spontaneous language production. This behaviour is illustrated in

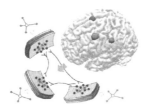 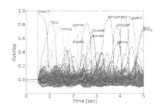

(a) Conceptual derivation of the Potts network

(b) An example of latching dynamics in the Potts network

Fig. 1: (a) Conceptual derivation of the Potts network, which models the cortex as a simplified two-level, local and global associative memory network. The local dynamics of a cortical patch (sheets) are reduced to a Potts unit with several states (here, 4). The global associative memory is the collection of the Potts units, which are connected through long range connections (black connections between cortical patches). (b) An example of latching dynamics in the Potts network. The x axis is time and the y axis is the correlation (overlap) of the network state with specific stored patterns. Each memory pattern represents a word. In this simulation the memory patterns are randomly correlated. [Adapted from Brovelli et al. (2002)][27].

Fig. 1(b), which shows the overlap between the actual network activation and the activation pattern that characterises the stored patterns as a function of time. Initially, an externally cued attractor leads to retrieval—a full retrieval corresponds to an overlap of one. However, the activation of the network does not remain in the retrieved pattern. Instead, it jumps or latches from attractor to attractor, driven by adaptation effects. Jumps between attractors are facilitated by an overlap between the current and the subsequent memory pattern.

4. Implementation of Word Representation in the Potts Network

After constructing the training language, BLISS, and implementing the Potts network, we need to represent the BLISS words into the network.

We represented the BLISS words in a distributed fashion on 900 Potts units, among which, 541 units express the semantic content (comprising the *semantic sub-network*), and the rest, 359 units, represent the syntactic characteristics of a word. The distinction between the semantic and syntactic characteristics of a word has been loosely inspired by neuropsychological studies [29,30]. We have also made a distinction between the representation of function words (e.g. prepositions, determiners, and auxiliary words) and content words (e.g. nouns, verbs, and adjectives), as suggested by several neuropsychological findings [31,32]. While the sparsity (the fraction of active Potts units) was kept the same for all the words ($a = 0.25$) on all 900 units, it is not equally distributed between semantic and syntactic sub-networks: semantic units are less active in the case of the function words (45 active units out of 541 units) compared to the content words (135 out of 541), whereas syntactic units are more active for the function words (180 out of 359) than for the content words (90 out of 359).

To have a word-generating algorithm that reflects the variable degree of correlation between words, we used an algorithm comprised of two steps [23]: (1) we establish a number of vectors called *factors* or features. Each factor influences the activation of some units in a word by "suggesting" a particular state to that unit. A word can be called a child, as it is generated by several factors as parents. (2) The competition among these factors through their *suggestion weights* determines the activation state of each unit in a word. In each unit, the state with the strongest suggestion is the winner. In order to maintain the desired level of sparsity, we picked the units with stronger suggestions in their selected states, and inactivated the remaining units by setting them to the null state.

To determine suggestion weights of a factor for its child, we used, whenever possible, the co-occurrence of the factor and its child in the BLISS corpus generated by *Subject-Verb* model. As we generate each word category in the next sections, we will specify our choice for the suggestion weights.

The algorithm includes a noise term to avoid generating words with very high correlation. We produced a number of additional factors, called *hidden* factors, whose suggestion weights were randomly selected from the distribution of the weights of the visible or main factors.

The proposed word-generating algorithm can be argued to be consistent with the findings of recent fMRI computational studies, which attempted to predict the neural signature of words by considering some other words as features—the factors in our algorithm. For instance, in [33], the fMRI neural representations of some nouns were predicted by proposing a linear model that considered 25 verbs as features. In this study, features compete through weights that correspond to the co-occurrence of the feature and the main noun in a natural language. To test the ability of the model for predicting words by having a more diverse range of features, they considered 1000 frequent words instead of 25 as the features; the model again succeeded in predicting the fMRI BOLD response of the nouns, though with lower accuracy.

Generating words using the above algorithm, we quantified the correlation between two words as the number of active units that are at the same state (Nas) in both patterns. We use the notation of $< Nas >$ to measure the average correlation across all words of two word categories—either the same or different categories.

If two patterns μ and ν are randomly correlated, we expect the correlation measure to be $Na^{\mu} \frac{a^{\nu}}{S}$, where a^{μ} is the sparsity of pattern μ and S the number of active states that a unit of this pattern may occupy. If we store randomly correlated patterns in the semantic sub-network ($N_{sem} = 541$), this measure reads: $Nas_{sem} \simeq 4.8$ between two words with $a = 0.25$, and $Nas_{sem} \simeq 1.5$ between two words with $a = 0.25$ and $a = 0.08$; and, $<Nas_{sem}> \simeq 4.1$, averaged over all words, if there are 134 words with $a = 0.25$ and 15 words with $a = 0.08$. On the other hand, with randomly correlated patterns in the syntactic sub-network ($N_{syn} = 359$), this measure reads: $Nas_{syn} \simeq 3.2$ between two words with $a = 0.25$, and $Nas_{syn} \simeq 6.4$ between two words with $a = 0.25$ and $a = 0.50$; and $<Nas_{syn}> \simeq 3.3$, averaged over all words, if there are 134 words with $a = 0.25$ and 15 words with $a = 0.50$.

4.1. *Semantic Representation*

To generate the semantic units of words, first nouns were generated using some *feature norms* as factors—feature norms include a list of features for a concept (e.g. *is-animal* and *a-mammal* are the features of *dog*). We then generated adjectives and verbs using nouns as factors. Finally, nouns, adjectives, and verbs served as factors for the generation of proper nouns and function words.

For the representation of nouns we used the feature norms in the McRae database [34]. The database was collected from an experimental study in which 541 nouns, including 18 BLISS nouns, were associated by human participants with a set of feature norms. In total, for all 541 nouns, 2500 features (e.g. *is-made-of-metal*, *is-animal*, *a-mammal*) were used; out of 2500 features, 190 features were associated to 18 BLISS nouns. We represented these features as vectors of 541 elements with $a = 0.25$.

To represent these 190 features as vectors with 541 elements we followed several steps: (1) we sorted the features, $f_1...f_i...f_{190}$, in descending order, by the number of concepts (or nouns) that are associated per feature, ω_{f_i}, in the database (e.g. $\omega_{an-animal} = 90$); (2) in an orderly fashion, we picked a feature in the list, randomly selected some units of its 541-element vector, then assigned their states by considering the previous features in the list as their factors. The number of randomly selected units was $3*(33+\omega_{f_i})$, because $<\omega_{f_i}> = 12$ and we needed to maintain the average sparsity around $a = 0.25$ (135 active units). As the first feature in the list did not have any preceding feature, we randomly assigned the states of its units. The suggestion weight was the co-occurrence frequency of features in the database; hence, the features that are more often associated with the same nouns will have higher correlation.

After the representation of the features of the McRae database, we used these 190 features as factors for the generation of the nouns. For a given noun, the features that are associated with that noun in the McRae database suggest the activation state of the units, with the weight of $\frac{1}{3*(33+\omega_{f_i})}$, to strengthen the uniqueness of the features. Hence the features that are more distinct in the database (e.g. *a-baby-cow* with $\omega = 1$) because of their smaller ω give more distinctive, stronger suggestions to a noun than popular features (e.g. *an-animal* with $\omega = 90$). The features that suggest the states of a noun and are associated with that noun in the database are likely to suggest other nouns that belong to the same semantic category; thus we expect higher correlations between words of the same semantic category.

After generating the semantic units of the nouns, we produced the semantic representation of the 37 verbs and the 18 adjectives of BLISS by using the nouns as factors. The suggestion weight of a noun for a verb or an adjective is determined by the co-occurrence probability of the noun and the corresponding word (either verb or adjective) in the BLISS corpus; hence the representation of a verb or an adjective tends to be more correlated with the nouns that appear more frequently with it in the corpus. For the generation of verbs and adjectives we added about 400 hidden factors in addition to their main factors, to avoid high correlations between these words. High correlations would have interfered with the dynamics of the semantic network.

After generating the semantic representation of nouns, verbs, and adjectives, we used these content words as factors—together with 400 hidden factors—to generate 6 proper nouns and then 15 function words.

As for the singular and plural form of words, we assumed that the meaning (the semantic part) should be the same for both numeral forms, and the only distinction should be in syntactic representation. Therefore, the plural and singular forms of nouns and verbs (e.g. *dog* and *dogs*, or *kill* and *kills*) are stored as identical in the semantic sub-network.

In the semantic representation, a generating factor influences the representation of a word by a weight that is proportional to the joint probability of the factor and the word. We have thus compared the correlation of some factors with the generated word in the Potts network and in the training BLISS corpus, generated by the Subject-Verb model (Fig. 2). This correlation was measured as $<Nas>$ in the Potts network, and as *joint probability* in the BLISS corpus (the joint probabilities between a word and its factors were normalized to 1). Although in the generation of the words, a very high noise level—about 400 hidden factors—was used to decrease the correlation between words, Fig. 2 demonstrates that a highly frequent word pair in the BLISS corpus still has a high correlation (*Nas*) in the semantic sub-network. These high *Nas* correlations indicate a deviation from the regime of randomly correlated patterns ($<Nas> \simeq 4.8$, between content words).

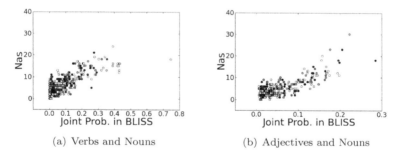

(a) Verbs and Nouns   (b) Adjectives and Nouns

Fig. 2: Comparison of the correlation between words and their factors in their semantic representation ($<Nas>$ on y-axis) versus their joint probabilities in the BLISS corpus produced by the Subject-Verb model (x-axis) (the joint probabilities between a word and its factors have been normalized to 1). Each dot indicates a pair of a word and its generating factor. (a) The correlation between verbs and nouns (the generating factors of verbs) in their semantic representation vs. the joint probability between the verbs and the nouns in the BLISS corpus (e.g. *kill sword*); Likewise, for (b) adjectives and nouns (adjectives' generating factors) (e.g. *bloody sword*).

4.2. *Syntactic Representation*

For the syntactic representation of words, we first generated function words using a limited set of somewhat arbitrarily designed syntactic features. Using function words as factors together with those syntactic features, we generated the syntactic representation of nouns, verbs, adjectives, and proper nouns.

As factors for generating the function words, we arbitrarily designed 19 syntactic features: 7 lexical categories (noun, verb, adj, conjunction, preposition, pronoun, adverb); 2 numbers (singular, plural); 1 negation; 3 determiners (indefinite, definite, properNoun); 2 locations (close, far); and 4 directions (from, towards, samePlace, above).

We represented the above syntactic features as vectors of 359 elements with $a = 0.25$ (90 active units), while keeping the representation of features within each of the above categories orthogonal to each other. For instance, for the first item, *lexical categories*: (1) we generated the representation of *lxc/noun*, by randomly selecting 90 units and arbitrarily assigning their activation states; (2) for *lxc/verb*, we activated the same units as in the *lxc/noun* but assigning different states; (3) for the rest of the members (i.e. *lxc/adj*, ...), we used the same procedure as in (2) while keeping all these features completely orthogonal. We took the same steps (1)–(3) for other categories listed above while keeping the features within a category uncorrelated.

Since these syntactic features will be used as the factors for all the words, we arbitrarily set their suggestion weights for the generation of different word categories, either function words or content words (see Table 1 for some examples).

We used the above 19 syntactic features, together with 20 hidden factors, as the factors for the syntactic representation of 15 function words. Using the function words and the syntactic features, together with 20 hidden factors, we generated the syntactic representation of 36 nouns (singular and plural), 74 verbs (singular and plural), 18 adjectives, and 6 proper nouns (singular and plural). The suggestion weights of the function words for the generation of a content word are determined by the joint probability of the two corresponding words in the BLISS corpus. Thus, if a content word has a higher co-occurrence with a function word in the corpus, the representations of these two words tend to be more correlated.

Generating the syntactic representation of all the words, we measured their correlations within and across different syntactic categories (singular and plural nouns, singular and plural verbs, adjectives, singular and plural proper nouns, and function words), as shown in Fig. 3. As expected, the correlations between relevant syntactic categories highly deviate from the regime of randomly correlated patterns in the syntactic sub-network; for randomly correlated patterns, $<Nas> \simeq 3.2$, between content words, and $<Nas> \simeq 6.4$, between content words and function words. As shown in Fig. 3(a), singular nouns (Nsg) have higher correlations with other noun categories (i.e. plural nouns and proper nouns) and also with other singular words (i.e. singular verbs), than with plural verbs or with adjectives. Though function words (Fwd) participate as factors in the generation of all the content words, their correlations with other categories are relatively small, even within the function words themselves, because of their high sparsity ($a_{syn}^{fwd} = 0.50$) compared to other words and to their syntactic features.

Using auto- and hetro-associative learning rules [34], we stored the implemented words into the network. Fig. 4 shows a preliminary result of interaction between the semantic (with randomly correlated patterns) and syntactic sub-networks. The detailed description of how we store the BLISS words and the grammar into the Potts network will appear elsewhere.

Table 1: Suggestion weights of some of the syntactic features, served as factors for the syntactic representation of words. Each element indicates the suggestion weight of a syntactic feature (labelled on the columns) for the generation of a word or words of a category (labelled on the rows).

	noun	verb	adj	conj	prep	pron	adv	sg	pl	neg	indf	def	propn
thatc	0	0	0	1	0	0	0	0	0	0	0	0	0
of	0	0	0	0	1	0	0	0	0	0	0	0	0
in	0	0	0	0	1	0	0.4	0	0	0	0	0	0
with	0	0	0	0	1	0	0	0.3	0	0	0	0	0
on	0	0	0.3	0	1	0	0.4	0	0	0	0	0	0
to	0	0	0	0	1	0	0.1	0	0	0	0	0	0
for	0	0	0	0.1	1	0	0	0	0	0	0	0	0
doesn't	0	1	0	0	0	0	0	1	0	1	0	0	0
don't	0	1	0	0	0	0	0	0.3	1	1	0	0	0
the	0	0	1	0	0	0	0	0.5	0.5	0	0	1	0
a	0	0	1	0	0	0	0	1	0	0	1	0	0
this	0	0	1	0	0	0.5	0	1	0	0	0	1	0
that	0	0	1	0	0	0.5	0	1	0	0	0	1	0
those	0	0	1	0	0	0.5	0	0	1	0	0	1	0
these	0	0	1	0	0	0.5	0	0	1	0	0	1	0
noun/sg	1	0	0	0	0	0	0	1	0	0	0	0	0
noun/pl	1	0	0	0	0	0	0	0	1	0	0	0	0
propn/sg	1	0	0	0	0	0	0	1	0	0	0	0	1
propn/pl	1	0	0	0	0	0	0	0	1	0	0	0	1
verb/sg	0	1	0	0	0	0	0	1	0	0	0	0	0
verb/pl	0	1	0	0	0	0	0	0	1	0	0	0	0
adjective	0	0	1	0	0	0	0.3	0	0	0	0	0	0

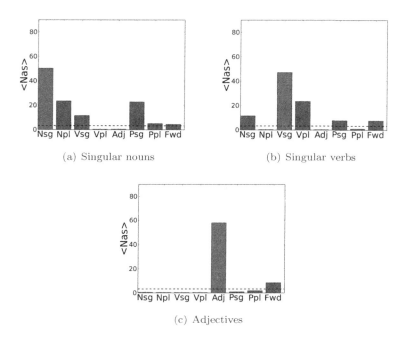

(a) Singular nouns   (b) Singular verbs

(c) Adjectives

Fig. 3: The average correlation $<Nas>$ of the syntactic representation of the words belonging to the same or different syntactic categories. (a) The correlation between the words that belong to singular nouns (Nsg) with themselves or with other word categories; likewise, for (b) singular verbs (Vsg), and (c) adjectives (Adj). The dashed lines indicate the expected correlations of randomly correlated patterns ($\simeq 3.2$, between content words).

5. Discussion

We have encoded words of BLISS, our artificial language of intermediate complexity, into a Potts attractor neural network, a simplified model of the cortex with large storage capacity that includes two components, semantic and syntactic.

The distinction between semantic and syntactic representations of words is inspired by neuropyschological findings [28,29]. We have also made a distinction between the encoding of function words and content words, as suggested by several studies [30,31]. While we keep the overall activity for these two categories the same over the network, semantic units are less active for the function words than for the content words, while syntactic units are more active for the function words than for the content words.

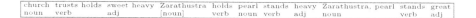

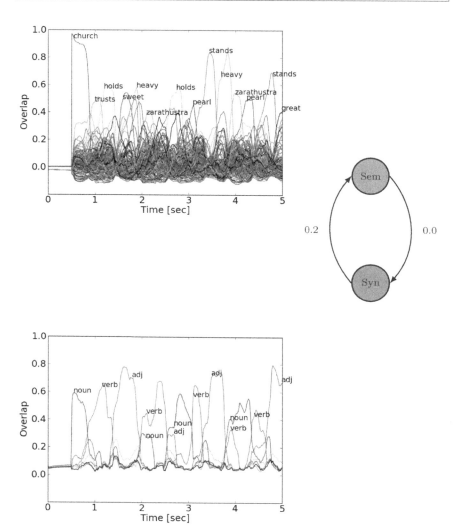

Fig. 4: The syntactic sub-network (bottom) influences the semantic sub-network (top) with the weight 0.2. The sentences produced by the interaction of these two sub-networks are written on the top. The parameters were set at $w = 1.6$, $\beta = 5$, $U = 0.0$; for the semantic sub-network, $C_{sem} = 100$, $\tau_{1\,sem} = 10$, $\tau_{2\,sem} = 200$, $\tau_{2\,sem} = 10000$; for the syntactic sub-network, $C_{syn} = 58$, $\tau_{1\,syn} = 5$, $\tau_{2\,syn} = 100$, $\tau_{3\,syn} = 5000$, $c_{auto} = 1.0$, and $c_{hetero} = 0.1$.

By distributing a word on a network, we stayed away from extreme localized approaches in which the sentence constructs are represented on distinct set of units [1,3]; on the other hand, by having a sparse representation of the words, which are implemented as a set of features localized on Potts units, we did not follow extreme distributed approaches [5]. Further, by making a distinction between semantic and syntactic characteristics of a word, we embedded grammar knowledge in the Potts network, unlike the case of a simple recurrent neural network [10].

In spite of the considerations we gave for word representation, there remain limitations and future questions that need to be answered. A word, beside semantic and syntactic properties, is also associated to a sound structure, a property that needs to be considered in future representations of the words. The current implementation of BLISS, the training language of the network, does not contain pronouns, interrogative sentences, or embedding structure. To investigate the ability of the network to produce such sentences, one needs to first examine the length of dependences that the Potts network can handle. For randomly correlated patterns, the sequences stretch beyond a first-order Markov chain [36]; however, this measure needs to be investigated with sentences generated by the semantic and syntactic sub-networks, given that these sub-networks can be trained with different statistics of word transitions derived from BLISS corpora generated by the different semantics models.

References

1. J. Hummel and K. Holyoak, *Psychological Review* **104**, 427 (1997).
2. C. R. Huyck, *Cognitive Neurodynamics* **3**, 317 (2009).
3. F. V. D. Velde and M. de Kamps, *Behavioral and Brain Sciences* **29**, 37 (2006).
4. T. Stewart and C. Eliasmith, Compositionality and biologically plausible models, in *The Oxford handbook of compositionality*, eds. M. Werning, W. Hinzen and M. Edouard (Oxford University Press, 2009)
5. R. W. Gayler, Vector symbolic architectures answer jackendoff's challenges for cognitive neuroscience, in *The Joint International Conference on Cognitive Science*, 2003.
6. T. Plate, Holographic reduced representations: Convolution algebra for compositional distributed representations, in *Proceedings of the 12th International Joint Conference on Artificial Intelligence*, 1991.
7. C. von der Malsburg, *Current opinion in neurobiology* **5**, 520 (1995).
8. P. Smolensky, *Artificial Intelligence* **46**, 159 (1990).
9. J. L. Elman, *Machine Learning* **7**, 195 (1991).
10. J. L. Elman, *Cognition* **48**, 71 (1993).
11. G. Borensztajn, The neural basis of structure in language: Bridging the gap

between symbolic and connectionist models of language processing, ph.d. thesis, Institute for Logic, Language and Computation, University of Amsterdam2011.

12. S. Pirmoradian and A. Treves, *Cognitive Computation* **3**, 539 (2011).

13. G. Marcus, S. Vijayan, S. Bandi Rao and P. Vishton, *Science* **283**, 77 (1999).

14. R. Gomez, *Trends in Cognitive Sciences* **4**, 178 (2000).

15. J.-R. Hochmann, A. D. Endress and J. Mehler, *Cognition* **115**, 444 (2010).

16. M. Ullman, R. Pancheva, T. Love, E. Yee, D. Swinney and G. Hickok, *Brain Language* **93**, 185 (2005).

17. R. de Diego-Balaguer, L. Fuentemilla and A. Rodriguez-Fornells, *Journal of Cognitive Neuroscience* **23**, 3105 (2011).

18. J. Bahlmann, R. I. Schubotz and A. D. Friederici, *Neuroimage* **42**, 525 (2008).

19. J. Lany and J. Saffran, *Psychological Science* **21**, 284 (2010).

20. D. Amit, *Modeling Brain Function: The World of Attractor Neural Networks* (Cambridge University Press, 1992).

21. J. Hopfield, *Proceedings of the National Academy of Sciences* **79**, 2554 (1982).

22. I. Kanter, *Physical Review A* **37**, 2739 (1988).

23. A. Treves, *Cognitive Neuropsychology* **22**, 276 (2005).

24. C. Fulvi Mari and A. Treves, *Biosystems* **48**, 47 (1998).

25. D. O'Kane and A. Treves, *Journal of Physics A: Mathematics and General* **25**, 5055 (1992).

26. V. Braitenberg and A. Schüz, *Anatomy of the Cortex: Statistics and Geometry.* (Springer-Verlag Publishing, 1991).

27. A. Brovelli, P. Battaglini, J. Naranjo and R. Budai, *Neuroimage* **16(1)**, pp. 130 (2002).

28. E. Kropff and A. Treves, *Natural Computing* **6**, 169 (2006).

29. T. Shallice and R. Cooper, *The Organisation of Mind* (OUP Oxford, 2011).

30. K. Shapiro and A. Caramazza, The organization of lexical knowledge in the brain: The grammatical dimension, in *Cognitive neurosciences*, ed. M. Gazzaniga (Cambridge, MA: MIT Press, 2004) pp. 803–814, 3rd edn.

31. A. Friederici and P. Schoenle, *Neuropsychologia* **18**, 11 (1980).

32. a. D. Friederici, B. Opitz and D. Y. von Cramon, *Cerebral Cortex* **10**, 698 (2000).

33. T. M. Mitchell, S. V. Shinkareva, A. Carlson, K.-M. Chang, V. L. Malave, R. a. Mason and M. A. Just, *Science* **320**, 1191 (2008).

34. K. McRae, G. S. Cree, M. S. Seidenberg and C. Mcnorgan, *Behavior Research Methods* **37**, 547 (2005).

35. H. Sompolinsky and I. Kanter, *Physical Review Letters* **57**, 2861 (1986).

36. E. Russo, S. Pirmoradian and A. Treves, Associative latching dynamics vs. syntax, in *Advances in Cognitive Neurodynamics (II): Proceedings of the Second International Conference on Cognitive Neurodynamics*, 2011.

UNEXPECTED PREDICTABILITY IN THE HAWAIIAN PASSIVE

ʻŌ. PARKER JONES*

Wolfson College, University of Oxford,
Oxford, United Kingdom
* E-mail: oiwi.parkerjones@wolfson.ox.ac.uk

J. MAYOR

FPSE, University of Geneva,
1211 Geneva 4, Switzerland
E-mail: julien.mayor@unige.ch

In Polynesian phonology and morphology, passives have traditionally been analyzed in two steps: (1) by assigning a single default pattern, which has been characterized as fully predictable and productive; and (2) by lexicalizing the remaining patterns, which have been assumed to be unpredictable. We overturn this assumption of unpredictability here for the Hawaiian passive by finding a computational model capable of predicting more than just the single default. This is achieved by learning statistical regularities over active verbs which predict their passive forms. Whether Hawaiian speakers are able to make use of these model predictions in a productive way can now be tested behaviorally.

Keywords: Hawaiian; Machine Learning; Morphology; Phonology; Polynesian Languages.

1. Introduction

Polynesian languages like Hawaiian signal the passive form of verbs by adding suffixes like *-hia* and *-lia*. Thus, for the active verb *inu* 'drink', the standard Hawaiian dictionary [1] lists two passive forms, each of which may be glossed as 'has drunk':

inu-hia
inu-mia

Although unlisted, a third passive form is implied since *ʻia* is completely productive:

inu ʻia

So there are really three passive forms for this verb; in Hawaiian, *'ia* is written as a separate word or particle, rather than as a suffix modifying a word on the view that productivity ought to confer orthographic independence [2]. Other words, however, take different passive suffixes. For example, *holo* 'run' takes *-kia*, as in *holo-kia* 'has run', while **inu-kia* is unattested (the prefixed asterisk denotes ungrammaticality). Following this logic, the following unattested *inu*-forms are all ruled out: **inu-kia* (cf. *holo-kia*); **inu-lia* (cf. *ka'a-lia*); **inu-nia* (cf. *kuhi-nia*); and **inu-ia* (cf. *mala-ia*). Table 1

Table 1. Some verbs and the passive suffixes they take[1]

Verb	Gloss	-hia	-kia	-lia	-mia	-nia	-ia	'ia
malu	'peace'	✓						✓
holo	'run'		✓					✓
ka'a	'roll'			✓				✓
inu	'drink'	✓			✓			✓
kuhi	'point'					✓		✓
mala	'tell'	✓	✓				✓	✓

illustrates the general situation, in which active verbs associate with different sets of passive suffixes, or types, and thus take different passive forms. (We ignore the *-a* and *-na* suffixes here for the sake of simplicity, noting that similar patterns posed no problem in a relevant previous analysis[10].) Taken together, these observations provide an outline of the puzzle: How does one know which passive forms occur with which verbs? We might also cast this puzzle in terms of prediction, in order to explore the inherent productivity of the system. How, then, does one *predict* the set of passive forms from active verbs in Hawaiian?

By asking the question in this way, however, we are already thinking beyond the received view of Polynesian passives [3,4]. In the received view, one, and only one, type of passive should be predictable. This results from asserting one passive type to be the default, a designation automatically available even when a verb has no other associated type. For example, the default in Hawaiian would be the particle *'ia*, which is found together with loanwords, as in *kalaiwa 'ia* 'has driven' (from English *drive*). To take another example, the default type varies dialectically in New Zealand Māori: On the North Island one finds the suffix *-tia*; on the South Island, *-ngia* [5]. However, in either dialect, one, and only one, default is assumed. In the received view, passive forms that do not take the default must therefore have, in one way or another, explicitly associated their passive type with the verb. Thus, that verbs take ʻ*ia* is expected to be predictable in Hawaiian, while taking -*hia* or -*lia*, or any other suffix, is not. Verb forms including these other types must be *lexicalized*. Interpreted as a cognitive theory, this suggests that, for each verb, speakers must effectively memorize every single passive form except for the default. By analogy, the view suggests that speakers somehow list passive forms in their mental lexicons as the Hawaiian dictionary lists passive forms with their associated active forms, the exception again being ʻ*ia*. As the default, ʻ*ia* is not explicitly listed in either mental lexicon or dictionary; it is rather accounted for by a grammatical rule. Underlying this view is the assumption that, but for the default, "a consonant of *unpredictable* quality appears in the passive ... forms" [5](emphasis added). In general, then, the received view has presupposed that passive forms are unpredictable.

In this chapter, our job is to demonstrate that this assumption is too strong: Multiple Hawaiian passives are in fact predictable, at a statistically reliable level of accuracy, from their associated active verbs. Our argument will take the form of an existence proof in that we shall produce a model capable of predicting more than one passive type, where the mere existence of this model contradicts received view. The model we describe is a feed-forward artificial neural network. Similar models have famously been employed in the literature to address the English past-tense [6–9]. We build here in particular on our previous neural network models of the Māori [10] and Hawaiian passives [11,12]. Our current model correctly predicts the traditional default, ʻ*ia*, while also predicting the types -*hia* and -*lia*, which represent the best attested of the suffix forms in our corpus. It is unexpected in the received view that these non-default forms should be predictable. What is more, we cannot rule out that with a bigger corpus the model might have predicted the other types too. A poignant correlate for endangered languages, like Hawaiian, is that changes in their vocabularies may affect the generalizations that they produce, as a limited corpus only allows us to predict a proper subset of passive types. We show through post-hoc analyses that our model relies on statistical distributions over the representations of active verbs to make its predictions, rather than on distributions over passive types alone. The model may thus be considered exemplar-based [13–15].

2. Data

The first part to the neural network simulation was to collect some data. We searched a machine-readable version of Hawaiian's standard reference dictionary [1], looking for entries containing the substring "Pas/imp." (for "Passive/imperative"). Here are a few example entries:

inu.hia. Pas/imp. of inu.

inu.mia. Pas/imp. of inu. (PPN inumia.)

This search returned 118 passive forms. But since we were interested in mappings from active verbs to *sets* of passive verbs, alternate passive forms (of the same active verb) were collapsed into passive morpheme sets. For instance, we found the passive forms *inuhia* and *inumia* in the dictionary, to which we added the default *inu 'ia*; these were then re-coded so that the verb *inu* mapped to the following set: {*-hia, -mia, 'ia*}. We also refer to such sets as passive classes. This reduced the original yield of 118 passive forms into 105 mappings from active verbs to passive classes.

Interestingly, one passive class turned out to be dominant; {*-hia, 'ia*} occurred a majority of the time (i.e., 62% = 65/105). We anticipated that this would lead to a problem when we interpreted the model [11]. The problem was that, in this case, the model could ignore the verbs and still achieve 62% accuracy. That is, if 65% of the dataset takes some class, then always guessing that a verb takes that class becomes a good blanket strategy for predicting passive forms, and the model would not need to learn to map different verbs to different classes.

We were able to sidestep this problem by boosting the dataset so that no class dominated. We did this by adding an extra 50, randomly-drawn verbs from the dictionary, each of which only took the default particle *'ia*. This yielded a new total of 155 mappings from active verbs to passive classes, where no class occurred more than 50% of the time. In particular, the problem case of {*-hia, 'ia*} occurred now only 42% (65/155) of the time. Moreover, this class remained below 50% (64/155) when one item was held out for testing the model (§4).

Table 2. Basic class to suffix correspondences

Class	Suffix
1	{*-hia*}
2	{*-kia*}
3	{*-lia*}
4	{*-mia*}
5	{*-nia*}
6	{*-ia*}
7	{*'ia*}

It will be convenient to use a shorthand when referring to the passive classes. It is unlocked by the key in Table 2. To represent the classes that contain more than one passive form, we will utilize a regular expression algebra where $+$ represents set union $\cup$[16]. For example, $1+7 = \{\text{-}hia\} \cup \{\text{'}ia\} = \{\text{-}hia, \text{'}ia\}$. Table 3 provides an overview of the passive dataset,

Table 3.   Overview of the passive dataset

Class	Suffixes	Count	Example
1+7	$\{\text{-}hia, \text{'}ia\}$	65	*wele*
7	$\{\text{'}ia\}$	50	*pala'ewa*
3+7	$\{\text{-}lia, \text{'}ia\}$	22	*pupu'u*
1+3+7	$\{\text{-}hia, \text{-}lia, \text{'}ia\}$	6	*wa'awa'a*
5+7	$\{\text{-}nia, \text{'}ia\}$	4	*wala*
2+7	$\{\text{-}kia, \text{'}ia\}$	2	*holo*
1+4+7	$\{\text{-}hia, \text{-}mia, \text{'}ia\}$	2	*inu*
4+7	$\{\text{-}mia, \text{'}ia\}$	1	*ho'ohau*
2+3+7	$\{\text{-}kia, \text{-}lia, \text{'}ia\}$	1	*kū*
1+3+5+7	$\{\text{-}hia, \text{-}lia, \text{-}nia, \text{'}ia\}$	1	*'aihue*
1+2+6+7	$\{\text{-}hia, \text{-}kia, \text{-}ia, \text{'}ia\}$	1	*mala*
	Total	155	

focusing on the passive classes and their counts. We assume that every verb takes *'ia*; so every class includes 7. The classes are further ordered by count, from most to least attested.

Unfortunately, there were not enough data to expect predictability from most of these classes. It was not possible for the model to both train on and test for a class when there was just one instance of it. For instance, one could train on the 1 instance of class 4+7 or one could test for the 1 instance of class 4+7, but not both. Even when the counts were higher than 1 in a class, there might not have been enough statistical power for the model to learn it. We limited our analysis then to the top three classes, each of which had more than ten items apiece.

The operational hypothesis follows:

Operational Hypothesis:
Passive verbs in classes 1+7 and 3+7 are statistically predictable from their corresponding active verbs, given a suitable training set of other words.

Confirming this hypothesis (to statistical significance) would cast doubt on the literature's assumption about thematic consonants. There were, incidentally, also enough data in class 7 to test its predictability. But this class only contained the default, which was not our primary interest to predict. When training the model, we included all of the classes, even the sparsely represented ones (with fewer than ten instances in the corpus). Since many of the classes are compositional, it is conceivable that these impoverished classes might contribute to the targeted classes. So, including them might help the model. It is also likely that including these classes adds noise to the system, making the model's task more difficult and its results more impressive.

Finally, while it is useful to speak of sets of *morphemes*, the system can always be interpreted as a mapping from active to passive verbs. For example, $inu \rightarrow \{$-*hia*, -*mia*, '*ia*$\}$ is shorthand for $inu \rightarrow inuhia$, $inu \rightarrow inumia$, $inu \rightarrow inu$ '*ia*. In this case, the output class becomes $X \rightarrow \{Xhia, Xmia, X$ '*ia*$\}$ for the input verb $X = inu$. This generalizes to all other cases as expected.

3. Methods

This section describes four crucial aspects of the neural network simulations. First, it describes how the data were pre-processed so as to be meaningful to the model. Second, it provides the model's architectural specifications. Third, it presents the error measures. Finally, it presents baseline measures which were used as a yardstick for the model's performance.

3.1. *Pre-processing*

Pre-processing involved translating verbs and passive classes into bit vectors. Since there were seven output classes, we modeled the output as a seven dimensional bit vector with the following key:

$$[\text{-hia -kia -lia -mia -nia -ia 'ia}]$$

where 1 meant that the suffix occurred, and 0 that it did not. To see how this works, consider a mapping like $inu \rightarrow \{$-*hia*, -*mia*, '*ia*$\}$. Its output vector received the following coding:

$$[1\ 0\ 0\ 1\ 0\ 0\ 1]$$

This illustrates the method for coding passive classes in the outputs vectors.

Coding active verbs in the input vectors was more complex. To include prosodic information in the model, the orthographic strings were parsed into syllables using a three-cell span. The first position in the span was reserved for the onset of the syllable (a consonant or an empty string). The second position was reserved for the syllable's main vowel. If the syllable was a diphthong, then the additional vowel was placed in the third position. In monophthongs, the third position contained the empty string.

We tailored the input vector for the longest verbs in the dataset. These were four syllables in length. Shorter verbs were right-aligned, so that spans to the left were filled with empty strings. This choice was motivated by locality, which is the idea that the phonological information at the end of the base, which is closer to the passive morphemes, should contribute more to the prediction of these morphemes than more distant phonological information nearer to the beginning of the base. Table 4 illustrates the scheme for each of the attested CVV patterns found in the dataset, where the symbol '–' designates the empty string.

The final step towards representing input verbs as bit vectors was to transform each phonological segment into a number. We did this by mapping each segment to a unique feature analysis. Typically, phonological membership into a feature class (like 'voiced') is indicated with +, while non-membership is indicated with −. Given that we used a bit encoding, 1 may be interpreted here as + and 0 as −. Incidentally, the feature analysis we used is not particularly significant; in a previous study, we found that this feature analysis outperforms a random encoding [11]. However, to make the model more interpretable, we have used the feature analysis rather than the random bit encoding. Table 5 shows the mappings that we used from segments to feature vectors. The features in Table 5 are descriptive features, which mix categories from the IPA and phonological literature. This analysis is also bigger than strictly necessary to define all of the segments we use. Whereas a 44-dimensional vector would have sufficed (i.e., (4 bit onsets + 4 bit primary vowels + 3 bit secondary vowels) × 4 syllables), we have used a 132-dimensional vector (i.e., (11 bit onsets + 11 bit primary vowels + 11 bit secondary vowels) × 4 syllables). The reason for this was pragmatic: Our primary aim was to find a model that worked. To do that we started with a richer phonological representation than was strictly necessary. Further work might of course improve upon this initial, ad hoc feature analysis, both in terms of cognitive reality and perhaps universality. For now, we merely note that the contrastive idea of a universal feature set is not without its problems [17]. Ad hoc feature analyses may be more cognitively realistic.

Table 4. Example mappings from aligned input strings to output vectors

Table 5. Mappings from segments to feature vectors

		labial	*lingual*	*glottal*	*nasal*	*plosive*	*continuant*	*vocalic*	*long*	*round*	*high*	*low*
h	→	0	0	1	0	0	1	0	0	0	0	0
k	→	0	1	0	0	1	0	0	0	0	0	0
l	→	0	1	0	0	0	1	0	0	0	0	0
m	→	1	0	0	1	0	0	0	0	0	0	0
n	→	0	1	0	1	0	0	0	0	0	0	0
p	→	1	0	0	0	1	0	0	0	0	0	0
w	→	1	0	0	0	0	1	0	0	0	0	0
'	→	0	0	1	0	1	0	0	0	0	0	0
a	→	0	0	0	0	0	0	1	0	0	0	1
ā	→	0	0	0	0	0	0	1	1	0	0	1
e	→	0	0	0	0	0	0	1	0	0	0	0
ē	→	0	0	0	0	0	0	1	1	0	0	0
i	→	0	0	0	0	0	0	1	0	0	1	0
ī	→	0	0	0	0	0	0	1	1	0	1	0
o	→	0	0	0	0	0	0	1	0	1	0	0
ō	→	0	0	0	0	0	0	1	1	1	0	0
u	→	0	0	0	0	0	0	1	0	1	1	0
ū	→	0	0	0	0	0	0	1	1	1	1	0
–	→	0	0	0	0	0	0	0	0	0	0	0

3.2. *The model*

For the model itself, we employed a fairly standard feed-forward neural network architecture with input, hidden, and output layers. The output layer contained seven units, for the seven passive classes. The input layer contained 132 units to account for the bit vector representation of active verbs, which we described above. In short, the transformation from segmental representation to bit vector used three cell spans to represent each CVV syllable, explicitly using empty strings for unfilled onsets and secondary vowels. The longest verbs were four syllables long, so we used four of these spans . Finally, each segment was converted into an 11-bit vector (Table 5).

The number of units in the hidden layer was optimized in a preliminary study, which was designed to minimize the mean squared error after training the network on a subset of data and then evaluating the result against a validation set. (An additional test set was held back during the preliminary study, only for use in the passive study.) Various numbers of hidden units were evaluated until the optimum number of twenty units was discovered.

Connectivity between layers was all-to-all with no lateral connections. Moreover, bias units were present in both the hidden and output layers.

A standard sigmoid activation function was used to compute unit activation. All weights were initialized with a random distribution centering around zero, within the range $[-0.1, 0.1]$. The network was trained using the back-propagation algorithm. The learning rate was set to 0.1 while a momentum term was set to 0.05. The model was trained for 500 epochs (52,500 sweeps), using a mass (rather than incremental) training set. This means that all stimuli were presented each epoch randomly and without replacement. Further details can be found elsewhere [12].

The model was evaluated using leave-one-out rotation estimation. Using this technique, we trained the model on all but one of the input–output mappings, and tested the held-out mapping. The model was then re-initialized with random weights. This was repeated so that every data point was tested, maximizing power of the dataset and providing for an item-by-item analysis. We turn now to the error and baseline measures we used to evaluate the model's performance.

3.3. *Error measures*

Two error measures were employed in this study: the Mean Squared Error (E) and the Classification Error (P^C).

3.3.1. *Mean Squared Error*

The Mean Squared Error E measures the average square of differences between outputs and their targets. This is a continuous error measure in Euclidean space, and can be defined for some input pattern i as follows

$$E_i = \frac{1}{7} \sum_{j=1}^{7} \left(output(i,j) - target(i,j) \right)^2 \qquad (1)$$

where i indexes an active verb and j indexes an output unit (representing a passive class). This shows how close the model has approximated the desired function, and can score the model's performance on each verb.

We also found it useful to define an error for each passive class. This did this by averaging the mean squared error for all verbs in a class, where the class C contained N active verbs:

$$E_C = \frac{1}{7} \sum_{j=1}^{7} \frac{1}{N} \sum_{i \in C} \left(output(i,j) - target(i,j) \right)^2 \qquad (2)$$

3.3.2. *Classification Error*

The Classification Error P^C measures how many output patterns of each class C the model fails to classify:

$$P^C = 100 - \left(\frac{100}{N} \sum_{j \in C} match(round(output(i,j)), target(i,j)) \right) \quad (3)$$

where

$$match(x,y) = \begin{cases} 1, & \text{if } x \text{ and } y \text{ match} \\ 0, & \text{otherwise} \end{cases} \quad (4)$$

and

$$round(x) = \begin{cases} 1, & \text{if } 0.5 \leq x \\ 0, & \text{if } x < 0.5 \end{cases} \quad (5)$$

We note that the *round* function applies to each of the output nodes independently. This means that it was possible to round an output to an unattested target, such as the output class containing all of the thematic consonants 1+2+3+4+5+6+7 (which did not occur in our dataset). In other words, we did not constrain this error measure on include only the actual target vectors that we wanted to find. P^C therefore represents a statistically conservative measure, more likely in this respect to underestimate the model's true performance.

3.4. *Baseline estimates*

We defined two baseline estimates and used these as yardsticks of the model's performance. The first of these was a random guess, which we adjusted to account for the known and unknown outputs. The second was a weighted guess. Unlike the neural network model, neither of these baselines had access to the input representations. At best, they could only exploit regularities in the output patterns. Therefore, they represented formalizations of the assumption that the thematic consonants are unpredictable from the active verbs.

3.4.1. *Random guess: adaptation to the range of target values*

The received view of Polynesian passives claims that the acceptable passive forms of any active verb are unpredictable. A simple way to model this is to assume that thematic consonants are distributed randomly with a flat distribution. What makes this simplistic is the assumption that each thematic consonant occurs as often as every other thematic consonant, as it is conceivable that some thematic consonants are more common than others. In Māori, for example, Sanders [18] found that two passive types (i.e., /-a/ and /-tia/) accounted for more than 70% of the data, while no other passive type accounted for more than 7% of the data. As outlined above (§2), the Hawaiian case is similar: The dominant 'ia type occurs with all verbs, while other types occur infrequently. So, we know that assuming a flat distribution over all types is too simplistic. Nonetheless, let us start by making this simplistic assumption, that all forms including 'ia occur equally often, and then build up to a less simplistic baseline.

To begin, then, we might know that there will be seven independent events, like coin tosses, but still be unable to predict the outcome of these events. For such a situation, the Random Error E_r can be defined as follows

$$E_r = \int_0^1 p(x)(1-x)^2 dx = \frac{1}{3} \tag{6}$$

assuming a flat distribution $p(x)$ between 0 and 1. This is just E to a binary output with random variables in the range of $[0, 1]$. According to this baseline, each outcome has a probability of $0.\bar{3}$.

E_r assumes that all output nodes take values of 0 or 1. But, as noted above, this sets up a straw-man: One of the target values is constant for all inputs since class 7 always returns 1. Moreover, the neural network model can safely be assumed to learn this constant output, by increasing the class 7 bias towards 1, while decreasing the weights on all other inputs to 0. In effect, this sets the unit to the bias value and ignores the inputs.

A fairer baseline, then, sets the error to 0 for class 7, as the model always predicts class 7. This gives the following Pseudo-Random Error E_{pr}:

$$E_{pr} = \frac{6E_r + 0}{7} = 0.286 \tag{7}$$

This is still like flipping seven coins, but now six of the coins are fair and one of the coins is biased. Each E_{pr} outcome has a probability of 0.286.

3.4.2. Weighted guess: adaptation to the distribution of target values

Even with the adjustment to account for the constant class, the random guess baseline assumes that the target values are as often 0 as 1. Yet, for each output unit, the distribution of these values is actually uneven. One might therefore sample the statistics of the target values to make a more informed guess while still ignoring the inputs, and this weighted guess can be derived as follows. Given both the proportion of target values p with

0 for a given unit j and the activation a_j, the optimal activation of a_j minimizes E:

$$E = pa_j^2 + (1 - p)(1 - a_j)^2 \tag{8}$$

A minimization with respect to the activation $\frac{dE}{da}$ gives the optimal activation value a_{dist} which also minimizes E:

$$a_{dist} = 1 - p \tag{9}$$

The value of E associated with this unit over all patterns finally defines the weighted guess baseline E_{dist}:

$$E_{dist} = p(1 - p) \tag{10}$$

Using the distribution of target values to optimize P^C is even simpler. For each output unit, just produce the most frequent value: 0 or 1. If the target values return 1 more than 50% of the time, then the best strategy is to output 1 by default. If it is the other way around, and the target returns 0 more than 50% of the time, then the best strategy is to output 0 by default. In the unlikely event that some target value is split in half between 0 and 1, this sub-case can be handled by random guess.

That the neural network model is capable of learning weighted guess by decreasing all of the input weights to 0 and then tuning the bias term to match the output distributions. On this strategy, the input makes no contribution to the model's predictions. It would be desirable to rule this contingency out, to aid the interpretation of our model's results. This was achieved preemptively by boosting the corpus with the 50 extra verbs, selected at random from Pukui and Elbert's dictionary (§2). These were verbs that only took class 7, and they reduce the distribution of the 1+7 class from 62% to 42% of the total number of patterns. With the exception of the default, none of the target values had a distribution above 50%. As a result, the weighted guess strategy only predicts the default. If the model contradicts this prediction, then it can be understood to have learned some statistical regularities over the inputs patterns (i.e., the active verbs), and not simply to have learned some version of weighted guess. The empirical question remains, however, as to whether the model did outperform the weighted guess.

4. Results

This section presents the results of the neural network simulations of the Hawaiian passive. These results are evaluated against the baseline measures of random guess and weighted guess, in order to answer the question of whether more than just the default passive can be predicted.

Table 6 summarizes the model's overall results, as compared to the random and weighted baselines. The model's overall mean squared error $(E_{network})$ was 0.083. This was lower than the random errors but higher than the weighted error (E_{dist}), which was 0.076. The overall results, however, obscured the specific contributions of classes 1+7 and 3+7.

Table 6.   Overall results

Measures	Results
Random Error (E_r)	0.333
Pseudo Random Error (E_{pr})	0.286
Distribution Error (E_{dist})	0.076
Network Error $(E_{network})$	0.083 ($\pm$0.094)

Table 7 therefore summarizes a per-class analysis of the model, as compared with the weighted guess. The random guess baselines are the same as above. Note that the network's error was generally higher for classes with lower counts. For classes 1+7 and 3+7, the model surpassed the random guess baselines. The model also surpassed the performance of E_{dist} for class 3+7, but not for class 1+7. Using these metrics, however, it was still hard to conclude much about the model's relative performance from these results.

Table 7.   Per-class results

Class	Count	Network Error (E_C)	Distribution Error (E_{dist})
1+7	65	0.061 ($\pm$0.026)	0.043
7	50	0.052 ($\pm$o.062)	0.039
3+7	22	0.090 ($\pm$0.130)	0.127
1+3+7	6	0.166	0.131
5+7	4	0.227	0.173
2+7	2	0.212	0.176
1+4+7	2	0.145	0.181
4+7	1	0.283	0.176
2+3+7	1	0.143	0.264
1+2+6+7	1	0.343	0.322
1+3+5+7	1	0.290	0.265

Table 8 therefore compares the model's classification performance with the baseline classification performance. In this case, the model easily outperformed the baseline, as P^C predicts none of class 1+7 or 3+7 correctly. By contrast, the model successfully predicted about 50% of each class. With this comparison, we gain confidence that the model has indeed learned to extract statistical regularities from the input.

Table 8. Classification success for classes 1+7 and 3+7, plus the default class 7

Class	Count	Network $(1 - E_C)$	Baseline $(1 - P^C)$
1+7	65	56.92%	0%
3+7	22	45.45%	0%
7	50	58%	100%

There is still some question of what the model has learned. Has it extracted statistical regularities in the input representations, as hypothesized? Or, has it learned to minimize the error, by optimizing the range of output values (an "output-based" strategy) without taking into consideration the input patterns? In order to rule out alternative output-oriented strategies, we used a Monte Carlo sampling method [11].

For classes 1+7 and 3+7, we ran a barrage of simulations exploring the effectiveness of P^C with increasing amounts of noise. This was meant to discover whether some purely output-oriented combination of weighted guess and chance could explain the model's performance. The simulations began with the noise set to 0, which was equivalent the P^C baseline itself. The noise was then gradually increased to 1, while the network performance (classification of patterns) was recorded. These results were finally compared against the model's performance, to see whether this performance could be explained by the output-based strategy.

Figure 1 illustrates the results of these simulations. The x and y axes list the number of patterns that were classified correctly in classes 1 and 3. The probability of the simulations predicting the model's results are given as p-values, represented in gray scale. The Figure denotes statistical significance with mid-gray ($\alpha = 0.05$). Darker shades are less likely and lighter shades are more likely. At best, the simulations classified about 40 of class 1 correct, but not more than 5 of class 3. We also see that there was a trade off in the simulations' performance between classes 1 and 3. The simulations cannot perform as well for both classes 1 and 3 as for either independently.

The neural network model's performance is indicated in Figure 1 by the white 'X' near the center. This shows that the model achieved a classification of 37 and 10 patterns, respectively, for classes 1 and 3. This is well away from the gray-to-white area in Figure 1; the model's performance lies clearly within the black area. This excludes the possibility that the model's performance can be explained by the output-based approach, without extracting statistical regularities in the input patterns ($p \ll 0.0001$).

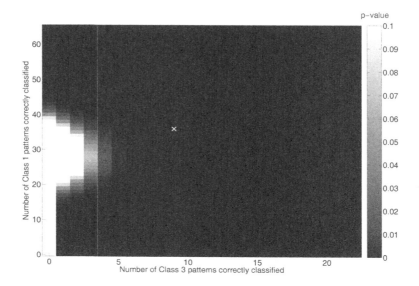

Fig. 1. Probability that the simulations of P^C with increasing amounts of noise do not predict the model's performance (see white 'X') on classes 1 and 3

To explore this result further, a Kolmogorov–Smirnov (K–S) test was applied on the activation levels of each output node when a pattern from the appropriate class was presented to the network versus when a pattern from another class was presented as an input. By doing so, we directly related the output produced by the network to the input. The prediction follows: If the network minimizes its output error without extracting statistical regularities from the input, there should be no statistical differences in the activation of an output node when patterns of different classes are presented to the network.

Table 9 shows that the activation levels of the output nodes 1 and 3 were statistically different when patterns from class 1 (respectively class 3) were presented to the network compared to when patterns from other classes were input to the network ($p \ll 0.0001$). Taken together, this supports an interpretation of the model in which it relies on input patterns (verb forms) to predict output classifications (passive forms). In sum, this contradicts the received view that all but the default passives are unpredictable in Hawaiian.

Table 9. *p*-values associated with a K-S test on the output node when patterns from a matching class are presented to the network and when patterns from another class are presented. Nodes 1 and 3 show meaningful differences, providing further support for the claim that statistical regularities in the input patterns help the network categorize patterns successfully.

Node	*p*-value
1	$\ll 0.0001$
2	0.36
3	$\ll 0.0001$
4	0.86
5	0.28
6	0.12
7	0.29

5. Conclusion

Polynesian passives have long fascinated linguists because of the presumed unpredictability of their patterns. The traditional solution to this puzzle has been to posit one passive type as a default, which is predictable and fully productive, while other forms have been assumed to be unpredictable [3,4]. In this chapter, we have overturned this assumption by showing that there are multiple predictable patterns in Hawaiian. In addition to the traditional default, *'ia*, the two most commonly attested suffixes in the corpus were also predictable, i.e., *-hia* and *-lia*. There were fewer than 6 tokens of any other passive type, which reflects its relatively limited documentation (e.g., compared to English). If more words took these other passive types, then our model might be trained to predict them too. Therefore, one direction for future work would be to train a similar model on token, rather than type, frequencies. In this case, even if few verbs take these other passive types, a larger set of token data might be enough to make them predictable in the model.

We also note in passing that our results carry implications for language change and grammar. If the words in a language affect the generalizations that speakers are able to make (e.g., *-hia* and *-lia* are predictable from our corpus, but the other suffix types are not), then changes to the corpus should imply changes to the grammar. (Here we conceive of a grammar generatively, as an ability to make productive generalizations about a language.) Equipped with the predictions of our model, the question can now be asked whether Hawaiian speakers can make productive use of these predictable patterns in the language. That is, we have found predictability within the system of Hawaiian passives, but we have not shown that speakers make use of this predictability. Further behavioral experimentation will be required for this.

As powerful as these results are, however, we also note certain limitations. For example, we do not propose that this model completely explains the Polynesian passive, even in Hawaiian. There are, of course, historical pressures at work, which may make use of additional perceptual mechanisms. The passive suffix *-ina* in Māori, for instance, seems to have developed by metathesis from *-nia*. Still, simple historical accounts alone cannot account for the language-specific innovations that we find across Polynesian languages either. Therefore, a combined explanation, including a diachronic account with our synchronic model, may prove more satisfying.

What is more, the predictability of multiple passive forms is not the only way in which our model challenges the received view. For example, the model presented here can be thought of as mapping surface active verbs to surface passive verbs [11,12]. Hence, the two levels of linguistic representation implied by both surface and underlying representations in the traditional generative approach turn out to be superfluous. Surface representations alone suffice. This is to say that our model may be interpreted as being *word-based* [19,20].

Also, as we showed through a range of analyses, our model relies on statistical distributions over the representations of active verbs to make its predictions. Hence, another way in which our model diverges from the traditional view is in the way it emphasizes rich, rather than impoverished, mental representations for words, and perhaps for phrases (such as verb + passive particle). Hence, our analysis may also be called *exemplar-based* [13-15].

Finally, the key innovation in our analysis has been the use of machine learning for drawing out subtle statistical patterns. Expanding upon previous work, which has successfully employed similar models to address morphological puzzles like the English past-tense [6,7], as well as Māori [10] and Hawaiian [11] passives, we showed neural networks here to be a useful tool for solving the puzzle of predicting multiple passive types from active verbs in Hawaiian.

Acknowledgements

This work was supported by a Mellon-Hawaiʻi fellowship (ʻŌPJ) and by Swiss National Science Foundation grant 131700 (JM).

References

1. M. K. Pukui and S. H. Elbert, *Hawaiian Dictionary* (University of Hawaiʻi Press, Honolulu, 1986).
2. W. H. Wilson, Manuscript. Attributed to the University Committee for the Preservation and Study of Hawaiian Language, Art, and Culture (1976).
3. K. Hale, *Journal of the Polynesian Society* **77**, 83 (1968).
4. K. Hale, *Current Trends in Linguistics 11*, ed. T. Sebeok (Mouton, The Hague, 1973) pp. 401–458.
5. J. Blevins, *Te Reo* **37**, 29 (1994).
6. D. Rumelhart and J. McClelland, *Parallel Distributed Processing: Explorations in the Microstructure of Cognition*, eds. J. McClelland, D. Rumelhart and the PDP Research Group (MIT Press, Cambridge, MA, 1986) pp. 194–248.
7. K. Plunkett and P. Juola, *Cognitive Science* **23**, 463 (1999).
8. K. Plunkett and V. Marchman, *Cognition* **48**, 21 (1993).
9. K. Plunkett and V. Marchman, *Cognition* **38**, 43 (1991).
10. ʻŌ. Parker Jones, *Proceedings of the Tenth Meeting of the ACL Special Interest Group on Computational Morphology and Phonology*, eds. J. Eisner and J. Heinz (2008).
11. ʻŌ. Parker Jones and J. Mayor, *Oxford University Working Papers in Linguistics, Philology & Phonetics: Papers in Phonetics and Computational Linguistics*, eds. ʻŌ. Parker Jones and E. Payne (2009).
12. ʻŌ. Parker Jones, PhD thesis, University of Oxford (2010).
13. J. L. McClelland and D. E. Rumelhart, *Journal of Experimental Psychology: General* **114**, 159 (1985).
14. K. Johnson, *Talker Variability in Speech Processing*, eds. K. Johnson and J. Mullenix (Academic Press, San Diego, 1997) pp. 146–165.
15. J. Pierrehumbert, *Frequency and the Emergence of Linguistic Structure*, eds. J. L. Bybee and P. Hopper (Benjamins, Amsterdam, 2001) pp. 137–157.
16. J. E. Hopcroft, R. Motwani and J. D. Ullman, *Introduction to Automata Theory, Languages, and Computation* (Addison-Wesley, 2006). 3rd Edition.
17. J. Mielke, *The Emergence of Distinctive Features* (Oxford University Press, Oxford, 2008).
18. G. Sanders, *Lingua* **80**, 149 (1990).
19. M. Haspelmath, *Understanding Morphology* (Arnold, London, 2002).
20. J. P. Blevins, *Journal of Linguistics* **42**, 531 (2006).

DIFFERENCE BETWEEN SPOKEN AND WRITTEN LANGUAGE BASED ON ZIPF'S LAW ANALYSIS

J.S. KIM

Biointelligence Lab, School of Computer Science and Engineering
Seoul, South Korea

C.Y. LEE

Interdisciplinary Program in Brain Science, Seoul National University
Seoul, South Korea

B.T. ZHANG

Biointelligence Lab, School of Computer Science and Engineering
Seoul, South Korea

Zipf's law states that the frequency of a word in a corpus is inversely proportional to its rank. The extension of Zipf's law to n-grams is an interesting and novel subject. We investigate the difference between spoken and written language based on an n-gram version of Zipf's law. N-grams are similar to n-th order hyperedges in a hypernetworks model. We find that clear and definite differences exist between the spoken and written language in 1- and 2-gram results. Also, the fitting parameter for the exponent of the rank varies between the spoken and the written corpora.

1. Introduction

If we sort each word in a text of a natural language in decreasing order of frequency, we find that the frequency falls off at an approximate rate of one over the rank. A text can be Wall Street Journal (WSJ), Wikipedia, and Vietnamese newspapers. This power law phenomenon is called Zipf's law. This power law appears widely in physics, biology, social science, computer science, earth and planetary science, finance, and economics.[1] The least effort principle was proposed as the origin of Zipf's law in human language.[2,3] Least effort for the hearer forces the speaker to generate a different signal for each object, at the expense of effort for the speaker. Multiple meanings of a word are required to satisfy the requirement of a hearer with a finite-size dictionary and memory. Because human memory is limited and objects are so numerous that polysemy naturally occurs. The power law was explained mathematically by the concept of inverse Fourier transforms[4], the broken stick[5], and Pitman–Yor process.[6] Moreover a power law distribution of the number of neighbors of each word was observed also in human language.[7] The human language network is inherently a small world network in that any word in the lexicon can, on average, be reached with fewer than three intermediating words.[7]

Zipf's law, can be explained by the Pareto principle. There are small numbers of the most frequent words, and the remainders are infrequent. Whether an n-gram of the corpus shows the power law or not is an interesting question. An n-gram is a sequence of n words and examples of a 3-gram (known as a trigram) are "how are you," and "by the way." The extension of Zipf's law to n-grams was shown in.[8] We found that an n-gram is similar to the n-th order hyperedge in hypernetwork model. A hyperedge is a higher-order extension of a traditional edge in node–node interaction in which the n-th order hyperedge is a set containing n nodes. If a weight is assigned to a hyperedge, its influence can be encoded in the hypernetwork.[9] We show that this n-gram analysis, i.e., n-th order hyperedge analysis, is an appropriate analysis for the investigation of the difference between spoken language and written language.

2. Methods

We analyze two kinds of document. The first is the script of TV drama Friends. All of the sentences spoken by the six main characters (Chandler, Joey, Ross, Monica, Phoebe, and Rachel) are collected into six files. The vocabulary size of each character is 2200 words in average. The second document is the Wall Street Journal text distributed by ACL/DCI. For this data, six corpora are chosen for analysis. The vocabulary size of each WSJ corpus is about 6200 words. A n-gram frequency vs. rank plot can be drawn through the following algorithm.

1. Make a dictionary of a corpus. The dictionary is then sorted in alphabetical order.
2. Slide a n words window through a sentence.
3. Write the index of each of the n words to an output file. The index is the position of that word in the dictionary made in step 1.
4. Sort the n output indices in increasing order.
5. Enumerate the number of times the same n indices occur in the sorted data from step 4.
6. Sort the enumerated numbers from the step 5 in decreasing order. The enumerated number is the frequency of that n-gram.
7. Plot a log–log frequency vs. rank figure and fit with a linear model for the 500 highest ranking words. (Log denotes the natural log.)
8. Find the words of the most frequent five n-grams.

3. Results

3.1. *Log– log frequency vs. rank plots*

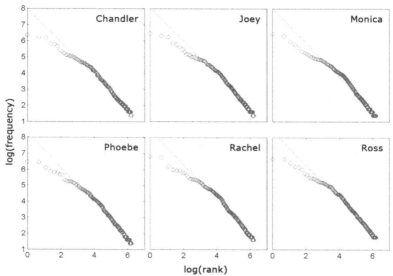

Fig. 1. 1-gram plot of Friends data

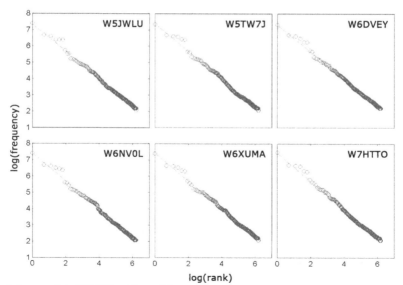

Fig. 2. 1-gram plot of Wall Street Journal data

In Fig. 1, the log(frequency) values are below the linear fit curve for log(rank) values between 0 and 3. However, the WSJ 1-gram result (Fig. 2) shows that log(frequency) values exhibit a good linear fit over this range.

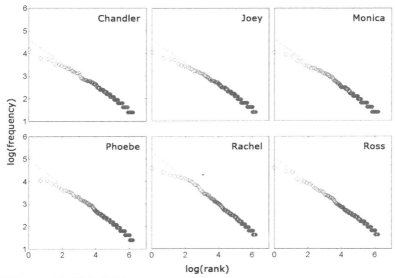

Fig. 3. 2-gram plot of Friends data

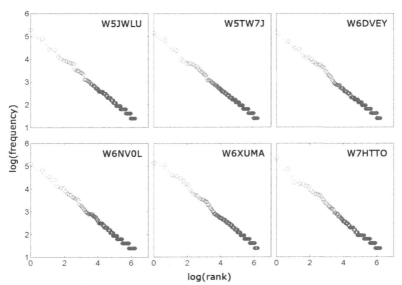

Fig. 4. 2-gram plot of Wall Street Journal data

In Fig. 3, the log(frequency) values are below the linear fit curve for log(rank) in the range from 0 to 2. Figure 4 shows that the WSJ 2-gram result again obeys this linear fit in the log(rank) range between 0 and 2.

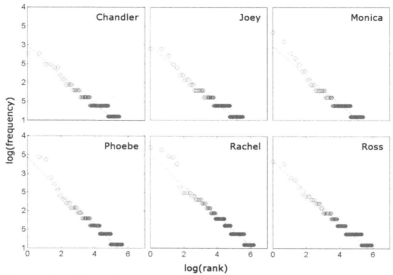

Fig. 5. 3-gram plot of Friends data

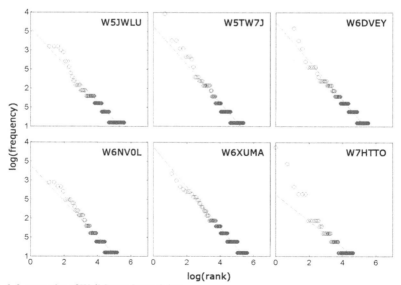

Fig. 6. 3-gram plot of Wall Street Journal data

Both Fig. 5 and Fig. 6 show log(frequency) values of the 3-gram data are above the linear fit curve in the log(rank) range of 0 and 2.

3.2. *Five most frequent words in 1-, 2-, and 3-grams*

Table 1. Top 5 most frequent n-gram words from the Friends data

	1-gram	2-gram	3-gram
Chandler	I, you, the, a, to	(you just),	(no no no),
		(in that),	(what are you),
		(this in),	(ew ew ew),
		(no know),	(I don't know),
		(I wanted)	(I can't believe)
Joey	you, I, the, to, a	(you kicking),	(I don't know),
		(I donated),	(no no no),
		(no my),	(you know what),
		(in that),	(what are you),
		(I wanted)	(oh my god)
Ross	you, I, the, a, to	(you know),	(I don't know),
		(no no),	(no no no),
		(join don't),	(you know what),
		(I mean),	(oh my god),
		(I just)	(what do you)
Monica	you, I, the, to, a	(you know),	(oh my god),
		(I mean),	(no no no),
		(this is),	(I can't believe),
		(no no),	(you know what),
		(I don't)	(why don't you)
Phoebe	you, I, the, and, a	(you know),	(la la la),
		(this is),	(oh my god),
		(I don't),	(you know what),
		(I mean),	(I don't know),
		(I know)	(no no no)
Rachel	I, you, the, to, oh	(I mean),	(oh my god),
		(you know),	(you know what),
		(are you),	(no no no),
		(this is),	(I don't know),
		(you guys)	(what are you)

In Table 1, "you" and "I" are dominant in the 1-grams, (I, mean) and (you, know) are frequent in the 2-grams, and (no, no, no) and (you, know, what) are common in the 3-grams.

68

Table 2. Top 5 most frequent n-grams in Wall Street Journal data

	1-gram	2-gram	3-gram
W5JWLU	the, of, to, a, in	(of the), (in the), (wall street), (street journal), (journal j)	(wall street journal), (street journal j), (mergers acquisitions tnm), (offers mergers acquisitions), (tender offers mergers)
W5TW7J	the, of, to, a, in	(of the), (in the), (wall street), (street journal), (journal j)	(wall street journal), (street journal j), (of the wall), (the wall street), (reporter of the)
W6DEVY	the, of, to, a, in	(of the), (in the), (wall street), (street journal), (journal j)	(wall street journal), (street journal j), (bank of new), (of new york), (the wall street)
W6NVOL	the, of, to, a, and	(of the), (in the), (wall street), (street journal), (journal j)	(wall street journal), (street journal j), (of the wall), (the wall street), (reporter of the)
W6XUMA	the, of, to, a, in	(of the), (in the), (wall street), (street journal), (for the)	(wall street journal), (street journal j), (cents a share), (in new york), (the wall street)
W7HTTO	the, of, to, a, and	(of the), (in the), (wall street), (street journal), (to the)	(wall street journal), (street journal j), (the wall street), (of the wall), (reporter of the)

Table 2 shows the results for the WSJ data. The dominant 1-gram words were "the," "of," "to," and "a." "You" and "I," which were dominant in the Friends script, do not appear in the written text. In the 2-grams, (of, the) and (street, journal) are dominant; (street, journal) is reasonable because the corpus is the WSJ data. (Wall, street, journal), (street, journal, j), and (reporter, of, the) are frequent 3-grams, which are again feasible in WSJ corpora.

3.3. *Exponent of rank*

Zipf's law can be stated as the following formula:

$$y = b / x^a,$$ (1)

where y is frequency, x is rank, a is the exponent, and b is a constant. Taking the logarithm of both sides of Eq. 1, we get:

$$\ln(y) = -a \cdot \ln(x) + \ln(b). \qquad (2)$$

Table 3. Exponent of the rank in Friends data with the top 500 ranked words

a	Chandler	Joey	Ross	Monica	Pheobe	Rachel	Mean	SD
1-gram	1.0328	1.0541	1.0593	1.0624	1.0899	1.0878	1.0644	0.0216
2-gram	0.5217	0.5117	0.5497	0.5334	0.5729	0.5927	0.5470	0.0311
3-gram	0.3759	0.3769	0.4106	0.3824	0.4475	0.4050	0.3997	0.0276

Table 3 lists the exponent values for each of the Friends' character's 1-, 2-, 3-grams. For the 1-grams, the exponent value is close to 1. The exponent value decreases as n increases. The mean value of a is 0.5470 for 2-grams and 0.3997 for 3-grams.

Table 4. Exponent of the rank in WSJ data with the top 500 ranked words

a	W5JWLU	W5TW7J	W6DEVY	W6NVOL	W6XUMA	W7HTTO	Mean	SD
1-gram	0.8358	0.8546	0.8329	0.8671	0.8525	0.8681	0.8518	0.0150
2-gram	0.6209	0.6026	0.6080	0.6132	0.6343	0.6108	0.5262	0.0764
3-gram	0.4828	0.5019	0.5139	0.4658	0.5307	0.4234	0.4864	0.0384

Table 4 gives the exponent values for the WSJ corpora. In the WSJ result, the exponent value a is 0.8518, which is smaller than the 1.0644 of the Friends data. The mean value of the exponent for the 2-gram is 0.5262, which is close to 0.5470 of the Friends 2-gram result, and the 3-gram exponent value 0.4864 is larger than the 0.3997 of the Friends 3-gram result. We can see the exponents of the rank between Friends script and WSJ is different in table 3 and 4.

4. Discussion

The log(frequency) values for the Friends 1- and 2-gram results were below the linear fit curve (Fig. 1 and Fig. 3). However, Fig. 2 and 4 show that the log(frequency) values were on the linear fit curve for the WSJ data. We suppose this result is due to the low frequency of the preposition and the determiner, such as "a" and "the" in the spoken language. In Table 1, we can see that "you,"

"I," and "the," are the most frequent words in Friends, and "the," "of," and "to" are the most frequent WSJ words. "You" and "I" are the most typical words in the Friends data. The frequencies of "you" and "I" in Friends data are smaller than those of "the" and "of" in the WSJ data. These results show clear differences between the spoken and the written language.

The exponents of the rank in the Friends 1-gram data are larger than those in the WSJ data. However, the 3-gram exponent shows the opposite trend. Tables 3 and Table 4 show this result. Further study on the exponent of the rank between the spoken and the written language is needed.

The frequency of an n-gram is the weight of the n-th order hyperedge in a hypernetwork. The weight of the n-th order hyperedge reflects the importance in the corpus. For a further study, n words selected with the allowance of jumping among words can give a new n-th order hyperedge. The traditional n-grams selected in this study were restricted to sequential words.

Zipf's law is studied in many research areas including music[10,11], cognitive science[12], mind[13], ingestion[14], city[15], and language evolution[16]. In addition, Pareto principle and the complex network theory are directly related concepts to Zipf's law. We hope our finding of the difference between spoken and written language can be extended to the cognitive linguistics based on the complex network theory in the future work.

Acknowledgments

This work was supported by the National Research Foundation of Korea (NRF) grant funded by the Korea government (MEST) (No. 2012-0005643, Videome/ No. 2012-0005801, BrainNet), supported in part by KEIT grant funded by the Korea government (MKE) (10035348, mLife), and in part by the BK21-IT program funded by MEST.

References

1. M. E. J. Newman, *Contemporary Physics* **46**, 323 (2005).
2. G. K. Zipf, *Human behavior and the principle of least effort: An introduction to human ecology*, (Addison-Wesley, Cambridge, MA, 1949).
3. R. F. I. Cancho and R. V. Solé, *PNAS* **100**, 788 (2003).
4. A. Czirók, H. E. Stanley and T. Viscek, *PRE* **53**, 6371 (1996).
5. R. Perline, *PRE* **54**, 220 (1996).
6. S. Goldwater, T. L. Griffiths and M. Johnson, *Advances in Neural Information Processing Systems* **18**, 459 (2006).
7. R. F. I. Cancho and R. V. Solé *Proc. R. Soc. Lond. B* **268**, 2261 (2001).

8. L. Q. Ha, P. Hanna, J. Ming and F. J. Smith, *Artif. Intell. Rev.* **32**, 101 (2009).

9. B. T. Zhang, *IEEE Compu. Intell. Magazine* **3**, 49 (2008).

10. M. Haro, J. Serrá, P. Herrera and Á. Corral, *PLos one* **7**, e33993 (2012).

11. B. Manaris, R. Juan and M. Penousal, *Computer Music Journal* **29**, 55 (2005).

12. C. T. Kello, G. D. A. Brown, R. F. I. Cancho, J. G. Holden, K. Linkenkaer-Hansen, T. Rhodes. and G. C. V. Orden, *TICS* **14**, 223 (2010).

13. G. Zamora-López, E. Russo, P. M. Gleiser, C. Zhou and J. Kurths, *Phil. Trans. R. Soc. A* **369**, 3730 (2011).

14. M. Myslobodsky and R. Coppola, *Rev. Neurosci.* **21**, 67 (2010).

15. Y. Chen, *Physica A* **391**, 3285 (2012).

16. A. M. Peterson, J. Tenenbaum, S, Havlin and E. Stanley, *Scientific Reports* **2**, 1 (2012).

READING ALOUD IS QUICKER THAN READING SILENTLY: A STUDY IN THE JAPANESE LANGUAGE DEMONSTRATING THE ENHANCEMENT OF COGNITIVE PROCESSING BY ACTION

HIRO-FUMI YANAI, TATSUKI KONNO*, ATSUSHI ENJYOJI[†]

*Department of Media and Telecommunications, Faculty of Engineering,
Ibaraki University, Hitachi, Ibaraki, Japan
E-mail: hfy@ieee.org*

As a simple demonstration of action and language integration in cognitive systems, we have compared the speeds of reading (saying) words in two experimental conditions. One condition was reading normally, i.e. reading aloud. The other was reading with your silent inner voice without making any sounds, i.e. your brain just simulates reading words without using your mouth, your tongue, nor your throat. We found reading aloud is significantly quicker than reading silently. This means that cognitive performance is facilitated with the existence of action to read.

Keywords: Reading Speed, Action-Language, Feedback, Cognitive Performance

1. Introduction

As a simple demonstration of action and language integration in cognitive systems, we have compared the speeds of reading (saying) words in two experimental conditions. One condition was reading normally, i.e. *reading aloud* or reading overtly. The other was reading with your silent inner voice without making any sounds, i.e. *reading silently* or reading covertly. Comparison of reading time for overt and covert conditions was first reported by Landauer (1962)[1], where reading times of numbers 1–10, 11–20, ..., 110–120, the English alphabet, and the American pledge of allegiance were recorded. And it was reported that the time required for recitations had no significant difference for overt and covert conditions. However there were claims that the results of Landauer (1962)[1] were not reproduced[2,3]. In a

---

*Currently with NTT DOCOMO, INC., Japan
†Currently with Hitachi Solutions, Ltd., Japan

study on the effect of mental practice on behavioral skill, Mackay (1981)[4] showed learning curves of times to produce sentences in mental and physical practice conditions, concluding that reading silently was faster than reading aloud.

Note that the meaning of the term *reading silently* used in this study is totally different from the meaning of the term in normal use of words. The normal meaning of the term is to read the sentences visually. Most of the time, when reading silently, we do not pronounce the words even internally. We do not even look at each letter of words, alternatively we ambiguously pattern recognize the words through our saccadic eye movements[5]. We use the term silently in a strict sense, i.e. by silently we mean that your brain just simulates reading words without using your mouth, your tongue, nor your throat, in other words, reading in the brain. In the silent condition, we have further restricted the participants in the way that they were requested not to use any part of the body. So that they were not allowed to nod, move their hands nor legs, and so on.

Now we would like to show how the present study is related to action-language investigation of cognitive systems. In Figure 1, a classical (old) view and a modern view (our view) of action-language relationship are compared. In Figure 1 (a), the Central Executive System processes and controls information related to action (A) and language (L), and A and L are independent. But in Figure 1 (b), apart from the Central Executive System, A and L could interacting with each other. This modern view is prevalent in recent literature (e.g., [6–8]. Traditional and modern perspectives of the links between action and language are reviewed together with the brain imaging data in Pulvermüller (2005)[9]. The purpose of our study is to test the hypothesis "A and L mutually interact with each other", in other words "the process P_{AL} exists", by means of behavioral data. We used mouth movement as a simplest example of A, and used word production as a simplest example of L. If the process P_L is impaired by inhibiting the process P_A, there exists interaction between A and L (the process P_{AL} exists). That is, action facilitates language.

In designing realistic and plausible systems, e.g. robots, the introduction of the process P_{AL} might help making the system efficient. Possibility is that interaction between A and L only through Central Executive System makes the whole system slower and heavier, but the existence of the process P_{AL} makes the system faster and lighter.

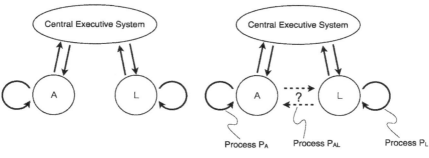

(a) Classical view of cognitive system   (b) Modern view (= our view) of cognitive system

Fig. 1.   A classical (a) and a modern (b) view of cognitive systems. *A*: Action, *L*: Language.

2. Material and Methods

Ten Japanese university students participated in the experiment. They are instructed to *read* single five-syllable words repeatedly as fast as possible until the experimenter stops them. To avoid possible interference of working memory loads with the task (e.g. counting the number of times they read), the experimenter stopped the task at random number of repetition times (repetition times were greater than or equal to ten). There were two conditions: reading aloud and reading silently.

We prepared four sets of words each of which was composed of eight words ($4 \times 8 = 32$ words in total). The four sets of words were selected as: a subset of Japanese-alphabetical syllables (JA), vowel-chart order syllables (VC), natural words (NW), and syllables randomized from NW (rNW, i.e. nonsense syllables). JA and VC were used to contrast the influences of tongue movement on the time required for reading. NW and rNW were used to contrast the influences of familiarity of words. Examples of words used are, "a-i-u-e-o" and "ka-ki-ku-ke-ko" (JA; see the second row of Figure 2 (a)), "a-e-i-u-o" and "ka-ke-ki-ku-ko" (VC), "u-ra-na-i-si" and "ri-so-u-te-ki" (NW), "si-u-ra-i-na" and "te-ri-u-ki-so" (rNW). The participants said each of 32 words in the two conditions (silently and aloud).

For the reader unfamiliar with Japanese alphabet, we would like to add some explanation on that. Current Japanese alphabet is usually aligned in two-dimensional table, with the consonant-axis and the vowel-axis. As can be seen by the table, each Japanese syllable is composed of a consonant and a vowel. Exception is the first row of the table in Figure 2 (a) that are composed only of vowels. Since fundamental part of the alphabet is

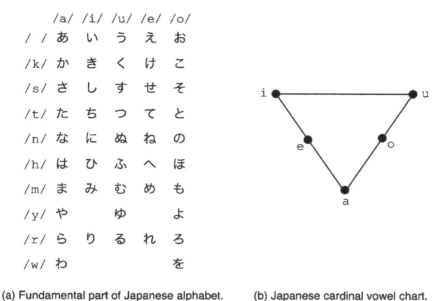

	/a/	/i/	/u/	/e/	/o/
/ /	あ	い	う	え	お
/k/	か	き	く	け	こ
/s/	さ	し	す	せ	そ
/t/	た	ち	つ	て	と
/n/	な	に	ぬ	ね	の
/h/	は	ひ	ふ	へ	ほ
/m/	ま	み	む	め	も
/y/	や		ゆ		よ
/r/	ら	り	る	れ	ろ
/w/	わ				を

(a) Fundamental part of Japanese alphabet.   (b) Japanese cardinal vowel chart.

Fig. 2. Japanese alphabet and a simplified diagram of cardinal Japanese vowels. (a) Japanese alphabet is almost always shown in two-dimentional arrays. Sometimes rows and columns are transposed. (b) Vowel chart based on tongue positions when pronounced.

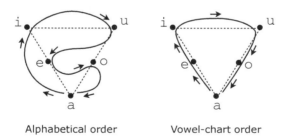

Alphabetical order   Vowel-chart order

Fig. 3. Alphabetical (JA) and vowel-chart (VC) order as a path on vowel chart.

composed of about 50 syllables it is called the table of the fifty syllables, some of the syllables having disappeared historically. All Japanese can say rows or columns of the table easily and automatically, and the table is used in every situation, from (paper) dictionary to the text input method of mobile phone. Therefore, although the task JA is, in the exact sense, not composed of natural words, it was as natural as NW for the participants to this experiment.

To investigate the effect of familiarity of words on the results, we used rearranged and randomized spelling versions of words (VC and rNW). In determining the sequences of VC, we followed the phonetic diagram of cardinal vowels by Daniel Jones[10]. The diagram represents the highest positions of the tongue when pronouncing the vowels. In the English version of the diagram of cardinal vowels, there are eight vowels. Since the Japanese language has only five vowels, the diagram could be simplified and schematized as Figure 2 (b). Since the sequence of JA, e.g. "a-i-u-e-o", does not follow the edges of the diagram, we added sequences which follow the clockwise path on the cardinal vowel chart, e.g. "a-e-i-u-o" (see Figure 3).

During reading, participants were requested to push a button in synchrony to their reading of each word. The time of each button push was recorded. This was to monitor their inner experiences that were not observable from outside. To make the mental and physical loads identical in the two conditions, they did button push also in the reading aloud condition.

3. Results

Relationship between the time required for *reading silently* and *reading aloud* are plotted for all of 320 samples in Figure 4. The result is composed of 10 participants × 4 task conditions (JA, VC, NW, rNW) × 8 words for each condition. From the figure we can see that reading aloud is quicker than reading silently. Time per word for two reading types in seconds are: Reading aloud (mean $= 0.63, \text{SD} = 0.07$) and reading silently (mean $= 0.73, \text{SD} = 0.13$). According to the analysis of variance (ANOVA), there is a significant effect of reading type on the time required ($F_{1,9} = 13.2, p < .01$). To look into the effect of word type on reading time, we have divided the data in Figure 4 into four figures for each of four word types (Figure 5). For each of four word types, ANOVA reads: JA ($F_{1,9} = 6.6, p < .05$), VC ($F_{1,9} = 10.9, p < .01$), NW ($F_{1,9} = 17.8, p < .01$), and rNW ($F_{1,9} = 16.2, p < .01$).

4. Discussion

As a simple demonstration of action and language integration in cognitive systems, we have shown that reading aloud is significantly quicker than reading silently. Surely this is a very simple experiment, but showing the facilitation of cognitive performance (syllable sequence generation in the present study) would be novel even in the field of embodied cognition.

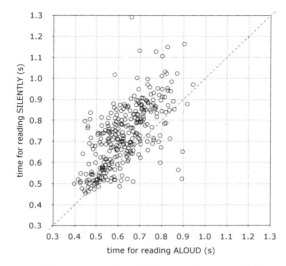

Fig. 4.   Time required for reading five-syllable words averaged for ten-time readings. Data points are for 10 participants for each of 4 word types, JA, VC, NW, rNW, and for 8 words, so that there are $10 \times 4 \times 8 = 320$ points.

To put it differently, the result says the simulation of the outer world (including the other parts of the body than the brain) is slower for the brain than doing in an embodied way. Some of the readers may agree that when in the dream, we cannot run as fast as we could actually do, and get irritated. This subjective experience would be related to the present result.

Our results are based on subjective reports from the participants. So that there might exist reliability problem because we do not have any tool to monitor the internal state of the brain. One thing we can say about the results is that the true time for reading silently is greater than or equal to the values stated above, i.e. the true speed of silent reading can be slower but scarcely not faster. This is to say the participants might unconsciously drop some of the syllables in silent reading. Significance level therefore might get even smaller.

Remaining problems include where the source of the difference lies. Or possible role of auditory feedback in this experiment. Distinction of effects by auditory feedback from motor feedback with reading would leads to better understanding of the present phenomenon. The experimental setup we might have to apply would be to cut off auditory feedback. Another problem is the learning effect. Although the task used here is simple, reading silently in the sense of this study is unusual for all of the participants. Preliminary

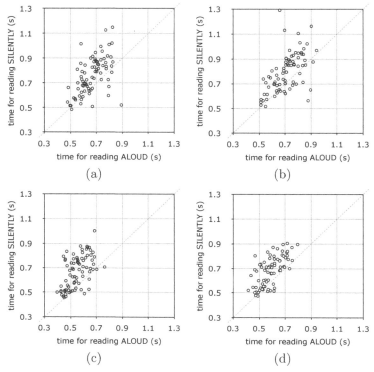

Fig. 5. Time required for reading words. The data in Figure 4 are divided into four word types, (a) Japanese alphabet (JA), (b) Japanese alphabet rearranged into vowel-chart order (VC), (c) Japanese natural words (NW), and (d) randomization of syllables of Japanese natural words (rNW).

results on the learning effect obtained from once-a-week experiment for a month show decrease in required time for two reading types, where the reading aloud remained quicker than silent reading. We are planning an experiment to look into the learning effect through a long span, e.g. for several months.

Finally we have to note on contradictory results of ours (reading aloud is faster than reading silently) to Landauer (1962)[1] (reading silently and reading aloud are equally faster), and Marshall and Cartwright[2,3], and Mackay [4] (reading silently is faster than reading aloud). We would like to show possible reasons of different results of ours, in particular, from Marshall and Cartwright[2,3] and Mackay[4]. One possible reason is the difference in instructions, i.e. how strong the demands of reading accurately and restricting movements were. Our demands were strong so that the participants tried not to skip any of syllables in silent conditions. Another possible reason is the difference in ratio of vowels and consonants in task words. Japanese language has a strong structure about syllables. Most of the words have the structure $cv - cv - \cdots - cv$, where c denotes consonant and v vowel. There is no syllable without v. Note that c could be NULL when the syllable is from the first row of the Figure 2 (a). Since consonants and vowels are processed differently (e.g., Ryalls, 1996[11]), we might have to discuss this sort of difference.

References

1. Landauer, T.K. (1962). Rate of implicit speech, Perceptual and Motor Skills, vol.15, p.646.
2. Marshall, P.H. and Cartwright, S.A. (1978). Failure to replicate a reported implicit/explicit speech equivalence, Perceptual and Motor Skills, vol.46, pp.1197–1198.
3. Marshall, P.H. and Cartwright, S.A. (1980). A final(?) note on implicit/explicit speech equivalence, Bulletin of the Psychonomic Society, vol.15, no.6, p.409.
4. Mackay, D.G. (1981). The problem of rehearsal or mental practice, Journal of Motor Behavior, vol.13, pp.274–285.
5. Starr, M. S. and Rayner, K. (2001). Eye movements during reading: some current controversies, Trends in Cognitive Sciences 5, 156–163.
6. Glenberg, A. M. and Kaschak, M. P. (2002). Grounding language in action, Psychonomic Bulletin & Review 9, 558–565.
7. Pecher, D. and Zwaan, R. (2005). Grounding Cognition (Cambridge University Press)
8. Glenberg, A. M., Sato, M., Cattaneo, L., Riggio, L., Palumbo, D., Buccino, G. (2008). Processing abstract language modulates motor system activity, The Quarterly Journal of Experimental Psychology 61, 905–919.
9. Pulvermüller, F. (2005). Brain mechanisms linking language and action, Nature Reviews Neuroscience, vol.6, pp.5760-582.
10. Jones, D. (1917). An English Pronouncing Dictionary (New York : E.P. Dutton).
11. Ryalls, J. (1996). A Basic Introduction to Speech Perception (Singular)

Development

TESTING A DYNAMIC NEURAL FIELD MODEL OF CHILDREN'S CATEGORY LABELLING

K. E. TWOMEY

University of Liverpool
Liverpool, UK.

J. S. HORST

University of Sussex
Sussex, UK

Recently, Dynamic Neural Field models have shed light on the flexible and dynamic processes underlying young children's emergent categorisation and word learning (DNF; e.g., Spencer & Schöner[1]). DNF models are a distinct class of neural network in which perceptual features can be represented topologically and time continuously, complementing existing connectionist models of cognitive development by building category representations that are available for inspection at any given stage in learning. Recent research in infant categorization and word learning has demonstrated that young children's ability to learn and generalise labels for novel object categories is profoundly affected by the perceptual variability of the to-be-learned category. We have captured these data in a DNF model of children's category label learning. Given a known vocabulary, our model exploits mutual exclusivity via simple associative processes to map novel labels to novel categories, and is able to retain and generalize these newly-formed mappings. The model was used to generate the testable prediction that children's generalizations of novel category labels should be contingent on the number and closeness of objects' perceptual neighbours. We present a replication of this prediction, via an empirical study with 30-month-old children. In line with the model, children were only able to generalize novel words to completely novel objects when those objects were central to the just-encountered category, rather than peripheral. This empirical replication demonstrates the predictive validity of DNF models when applied to cognitive development. Further, the data suggest that children's ability to categorise and learn labels is not a conceptually-based, stepwise phenomenon, but rather a graded, emergent process. As such, these data add weight to associative, dynamic systems approaches to understanding language learning, categorisation, and cognition more generally.

1. Introduction

The puzzle of how young children learn to categorise and label objects is well-known. Born into an enormously rich perceptual environment, from an early age children parse objects into categories and treat exemplars from a single category equivalently. By 18 months, children have begun to label these categories (e.g., Houston-Price, Plunkett, & Harris[2]), reliably inferring the referents of novel words despite the proliferation of potential referents.[3] This ability to form a quick, initial hypothesis about a word's meaning is known as fast mapping.[4] Several theoretical accounts of categorisation and word learning have been offered, from low-level associative learning (e.g., Smith[5]) to *a priori* conceptual primitives (e.g., Carey[6]).

Word learning and fast mapping have latterly been the focus of a variety of computational models (e.g., Colunga & Smith[7]; Gliozzi, Mayor, Hu & Plunkett[8]; McMurray, Horst & Samuelson[9]; Samuelson, Smith, Perry & Spencer[10]). Although inspired by an abundance of experimental literature, these simulations go a step further: unlike cognitive change in children, changes in a model's cognitive structure can be observed as they develop over time. Thus, computational models have made novel predictions about the cognitive structures underlying a variety of behaviours, across development.[11]

Importantly, however, these predictions must be empirically tested if a model's explanation for a behavior is to be taken seriously. The current paper presents just such a test. Twomey & Horst[12] describe a Dynamic Neural Field model (for an invaluable introduction and review see Spencer, Thomas, & McClelland[13]) which has successfully replicated data from an empirical study examining the effect of variability on 30-month-old children's category label learning. The current paper presents a novel prediction generated by the model (Simulation) and an empirical replication of that prediction (Experiment).

2. Simulation

2.1. *Dynamic Neural Fields*

Dynamic Neural Fields (DNFs[1]) are emergentist models of changes in neural activation in response to external stimuli. In contrast to their connectionist cousins (e.g., McClelland, et al.[14]), DNFs simulate neural structure and time continuously, with representations distributed across fields. DNFs are topologically functional, such that similarity on a given metric is represented by distance on a given axis.

Consisting of one or more input fields, DNF models initially receive input in the form of a modeller-defined increase in activation at a certain location in the field. These inputs represent responses to stimuli. Over time, the dynamics of the DNF allow peaks of activation to emerge in the thanks to locally-excitatory and both locally- and globally-inhibitory neural interactions; that is, activation spreads from a given location to its neighbours, whilst activation at more distant locations is suppressed. These peaks are taken to represent associations between stimuli.

Input fields are coupled reciprocally to memory fields, which are functionally similar to Hebbian weight changes in connectionist models. When a peak forms at a given location in the input field, activation spreads to the Hebbian field, where it decays slowly. The Hebbian field therefore helps new peaks form at the locations of previous peaks, simulating learning of associations over time.

2.2. Categorisation by Shared Features

Existing empirical and behavioural research demonstrates that children can categorise based on co-occurrence of perceptual features such as shape, colour and texture.[12,15,16,17] More recently, connectionist models have simulated the developmental differentiation of children's categories based on the assumption that categories are scaffolded from coherent covariation of perceptual features.[18] Taken together, evidence from empirical and computational studies suggests that exemplars that share perceptual features are perceived as perceptually similar by young children (see also, Sloutsky & Fisher[19,20]).

That children can extend known category labels to novel exemplars is not in dispute.[21,22,23,24] However, the number of features shared between even perceptually very similar items seems likely to affect label extension. We hypothesised that when taught a novel label for a novel category in a simulated fast mapping task, our model would generalise that label to new category members that shared many – but not few – features with previously-encountered exemplars.

2.3. Method

2.3.1. Architecture

The model is an adaptation of Faubel & Schöner's[25] simulation of dynamic feature binding and consists of a two-dimensional input field, representing perceptual similarity on one axis and labels on the other, as depicted in Figure 1 (see also Twomey & Horst[12]). Simultaneous presentation of inputs along the label and perceptual similarity axes may give rise to a peak at their intersection, representing an association between these two inputs—in behavioural terms, an association between a label and an object.

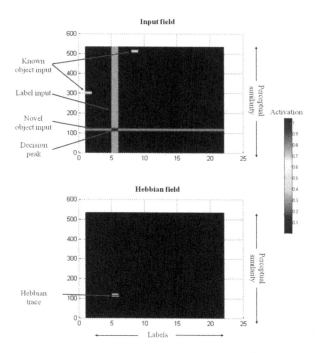

Fig. 1. Architecture of the Dynamic Neural Field model.

Because activation from these peaks is stored in the Hebbian field, during the familiarisation phase the model learns which labels are associated with which objects.

2.3.2. *Stimuli*

"Novel" object stimuli consist of input ridges along the perceptual similarity axis, which are generic along the label axis; that is, on the first presentation of a novel object, a peak can form at the intersection of that input and any location on the label axis, reflecting the fact that children in fast mapping tasks do not know the name of the novel objects they encounter. "Known" object stimuli consist of peaks of activation at locations in the input field representing the previously learned intersection between a label and an object. Label stimuli consisted of a ridge of activation along the label axis and could therefore be associated with any position along the feature axis. Exact locations of object inputs along the feature and label axes are given in Table 1.

Table 1. Locations along feature and label axes of inputs to the model. Inputs representing extension exemplars with many shared features are closer to exemplars seen during referent selection than inputs representing extension exemplars with few shared features.

	Referent Selection						Extension			
	Known object 1		Known object 2		Novel category exemplars		*Few*		*Many*	
	Object	Label	Object	Label	Object	Label	Object	Label	Object	Label
Category "*hux*"	40	21	470	13	113	5	119	5	114	5
	30	1	510	8	115	5				
	450	16	230	7	117	5				
Category "*cheem*"	150	9	60	4	263	11	269	11	264	11
	80	18	360	2	265	11				
	380	15	170	14	267	11				
Category "*doff*"	210	3	320	12	413	17	419	17	414	17
	340	20	20	6	415	17				
	490	19	190	10	419	17				

2.3.3. Design and Procedure

Our model simulates the empirical fast mapping paradigm, in which children are presented with multiple referent selection trials consisting of an array of two known objects and one novel object, and asked to retrieve the novel object in response to the novel label ("Can you show me the *blicket?*"; e.g., Horst & Samuelson[26]). The specific design of the empirical task discussed here is described in detail in section 1.2.1.3. In line with this empirical task, the model is presented with an initial referent selection phase, during which it is familiarised with three novel categories (each consisting of three exemplars) and three novel labels, presented in blocks of six trials per category. Each block consists of three known and three novel trials. Each novel exemplar serves once as the target (on a novel trial), and once as a competitor (on a known trial). The model is therefore presented with a total of 18 referent selection trials.

A single referent selection trial consists of an initial presentation of two known object peaks and a single novel object ridge. Then, the model is given a label input ridge, reflecting the experimenter's request for the novel object. This ridge crosses either a known object location (on known trials) or the novel object location and a known object location (on novel trials). A peak of activation may develop at one of these intersections, simulating the child's choice, which may or may not be correct.

Immediately following referent selection the model is given five trials in which no stimuli are presented, reflecting the delay between referent selection and test in the empirical task, and allowing the stored memory traces to decay.

Importantly, however, these memory traces do not decay entirely during this delay. Any remaining activation in the memory field may therefore influence peak formation (or "object choice") during the test trials. Following the delay the model receives three retention trials, identical to referent selection, except that object stimuli consist of three novel object ridges, one from each previously-encountered category. The model receives a different, previously-encountered novel label on each retention trial. Thus, the model can only accurately respond if the memory trace associating novel objects to novel labels is sufficiently robust. Finally, the model is presented with three extension trials to test label generalisation. At this stage, the model receives completely novel stimuli that share either *many* or *few* features with just-encountered exemplars. The model was run 24 times per condition.

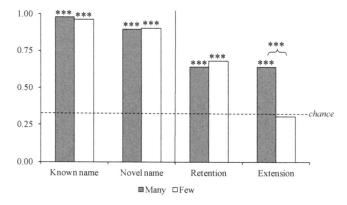

Fig. 2. Simulation results. *** $p < .001$. Chance = 0.33, all tests two-tailed.

2.3.4. *Results and Discussion*

Results from the simulation are depicted in Figure 2. During referent selection the model mapped known and novel labels to the correct referent at levels greater than expected by chance (0.33, all $ps < .001$, all t-tests two-tailed). At test, the model retained novel labels at above-chance levels in both conditions, *many:* $t(23) = 5.46$, $p < .001$, $d = 1.12$, *few:* $t(23) = 4.76$, $p < .001$, $d = 0.97$ (note that no difference was anticipated between conditions for retention, as stimuli presented during referent selection and retention are identical across conditions). In contrast, however, the model extended novel labels in the *many* condition, $t(23) = 5.44$, $p < .001$, $d = 1.12$, but did not extend novel labels in the *few* condition, $t(23) = -0.45$, *ns.*, $d = -0.09$. An independent samples t-test confirmed a significant difference between conditions for extension, $t(46) = 4.17$, $p < .001$,

$d = 1.23$. Thus, as predicted, the model generalised novel names only to objects that shared many features with the categories encountered during referent selection.

3. Experiment

Using the same architecture and procedure as a previous, successful DNF simulation of 30-month-old children's behaviour in a fast mapping task (Twomey & Horst, 2011), the DNF model predicts that children will extend previously fast-mapped novel names to completely novel exemplars that share many – but not few – features with previously-seen novel exemplars. The current experiment tests this prediction empirically with 30-month-old children using a design identical to the model. Importantly, the stimuli used during referent selection were identical across conditions until the extension trials when children were presented with exemplars that shared either many or few shared features with just-seen novel exemplars.

3.1. *Method*

3.1.1. *Participants*

40 typically developing, monolingual, English-speaking 30-month-old children (23 girls, $M = 29m$, 0d, $SD = 43.34d$; range = 24m, 11d - 32m, 17d) with a mean productive vocabulary of 521 words ($SD = 128.92$ words, range = 263 - 662 words) and no family history of colourblindness participated. Half of the children were randomly assigned to the *many shared features* condition, and the other half were randomly assigned to the *few shared features* condition. Children's ages and productive vocabularies did not differ between conditions. Data from 10 additional children were excluded from analyses due to fussiness (7), experimenter error (2) and illness (1). Parents were reimbursed for travel expenses and children received a small gift for participating.

3.1.2. *Stimuli*

Known objects consisted of eighteen toys from categories familiar to 2-year-old children, for example a plastic toy apple and a metal toy bus. Novel objects are depicted in Figure 3 and consisted of fifteen toys from three categories not familiar to 2-year-old children. Novel exemplars from a given category shared basic shape but differed in overall number of shared features, based on evidence that preschool children can differentiate shape components in 3D objects (or "geons"[27,28]) and categorise solid objects on the basis of shared shape.[22,29,30] Thus, within each novel category, exemplars shared more or fewer perceptual features (geons, colour) with other exemplars of that category. Test objects, depicted in the final two columns of Figure 3, were designed to share either *many* or *few* features with objects encountered during referent selection. To ensure that the *many-* and *few-shared-features* test objects were indeed appropriately similar or dissimilar to the objects encountered during referent selection, we conducted a Multidimensional Scaling Analysis (MSA; for discussion see Abecassis *et al.*[27]), calculating object similarity based on shared colour, geons and labels. The MSA confirmed that the *many* objects were more similar to the referent selection objects than the *few* objects[a]. Finally, novel category labels were the arbitrarily assigned nonsense words *hux, doff* and *cheem* (see also Twomey & Horst[12]).

3.1.3. *Procedure and Design*

Before the experiment began the parent was asked to complete a vocabulary checklist.[31] Parents were also shown colour photographs of all stimuli to ensure that they were appropriately familiar or novel. All children were familiar with all known objects, and no children were familiar with any of the novel objects.

| Label | Referent Selection | Extension | |
		Few	Many
hux			
doff			
cheem			

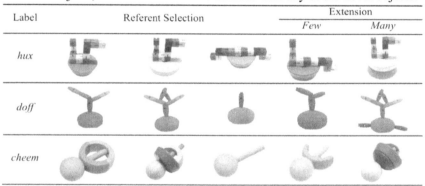

Fig. 3. Novel objects used in the empirical study.

The experiment began with three warm-up trials to familiarise children with the task. Stimuli were presented on a transparent plastic tray divided into three equal sections. Three known objects, chosen at random from the known objects

---

[a] To further confirm that the stimuli presented to the children reflected the stimuli presented to the model, 20 adults from the university community were asked to rate each object for similarity to the object shown in the second column of Figure 3 on an 11-point Likert scale. Scores reflected the distances between the model stimuli, and a subsequent re-run of the simulation with stimuli positioned at the exact locations dictated by these similarity scores generated the same pattern of results.

used during the referent selection phase, were presented to the child on the tray in pseudorandomly-determined position (i.e., left, middle or right). First, the experimenter held the tray stationary on the table and silently counted for three seconds to allow the child to look at the objects (see Horst & Samuelson[26]). Then, the experimenter asked the child to select one of the objects ("Which one's the cow? Can you show me the cow?"). All objects were labelled twice, with up to two more labelling instances when children needed encouragement. No object was labelled more than four times. The experimenter then slid the tray towards the child and allowed the child to point to or hand her one of the objects. Children were heavily praised for correct responses, and prompted to choose again for incorrect responses.

Referent selection trials immediately followed warm-up trials and proceeded in an identical manner, except that children were given no feedback following their choices. Each child encountered three novel categories (each consisting of three exemplars) and three novel labels, presented in blocks of six trials per category. Each block consisted of three known and three novel trials. Each novel exemplar served once as the target (on a novel trial), and once as a competitor (on a known trial). The children were therefore presented with a total of 18 referent selection trials, followed immediately by a five-minute delay.

After the delay, children were presented with a new warm-up trial to re-engage them with the task. Three retention trials immediately followed the warm-up trial and were identical across conditions. Retention trials proceeded in an identical manner to referent selection trials, except that children were presented with three novel exemplars on each trial: one previously-encountered exemplar from each novel category. Extension trials proceeded in an identical manner to the retention trials. Children were presented with three completely novel exemplars, one from each novel category. In the *many* condition, children were presented with the exemplars that shared many features with those encountered during referent selection. In the *few* condition, children were presented with the exemplars that shared few features with those encountered during referent selection.

3.2. *Results and Discussion*

Results from the empirical study are depicted in Figure 4. During referent selection children mapped both known and novel labels to the correct referent at levels greater than expected by chance (0.33; both $ps <.001$, all t-tests 2-tailed). At test, children retained novel labels at above-chance levels in both conditions, *many shared features:* $t(19) = 2.82$, $p <.05$, $d = 1.29$, *few shared features:* $t(19) = 2.20$, $p <.05$, $d = 1.01$. In contrast, children extended novel labels in the *many* condition, $t(19) = 3.74$, $p = .001$, $d = 1.72$; but did not extend novel labels in the *few* condition, $t(19) = 0.88$, *ns.*, $d = 0.31$. An independent samples *t*-test of extension between conditions approached significance, $t(38) = 1.85$, $p = .071$, $d = 0.60$. Thus, children's overall pattern of responding reflected the overall pattern generated by the model.

These data support "correlated features" accounts of categorisation, for example the classic Younger & Cohen studies.[15,32] In these studies, 10-month-old infants were sensitive to correlations between configural and perceptual attributes in novel 2D animal stimuli (see also Plunkett, Hu and Cohen[33]; Rakison and Cohen[34]; Younger and Cohen[32]; Younger, Hollich, and Furrer[35]). The current study demonstrates that older children can also generalize labels systematically based on correlations between perceptual features such as geons and colour. Importantly, the empirical data replicate the simulated data, indicating that the DNF model constitutes an informative simulation of young children's category label learning.

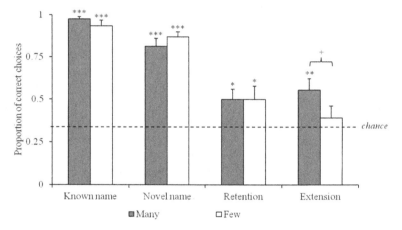

Fig. 4. Children's proportion of correct choices. Dotted line represents chance (.33). Error bars represent one standard error. *** $p < .001$, ** $p .001$, * $p < .05$, + $p = .071$.

4. General Discussion

This paper presents an experimental replication of predictions generated by a computational model of young children's word learning and categorisation. Based on a Dynamic Neural Field model of children's word learning via mutual exclusivity[12], we simulated children's behaviour in a fast mapping task to examine the nature of children's noun extensions after familiarisation with an object category. The model predicted that children would extend labels only to novel exemplars that shared many features with the familiarised category, and these predictions were borne out empirically.

In line with earlier applications of DNFs to developmental phenomena such as the A-not-B error[36], spatial binding of objects to labels[10] and the shape bias[37], this model successfully simulates apparently complex behaviour using simple low-level associative processes. Importantly, during referent selection the model maps novel words to novel referents without any preprogrammed "reasoning" ability. Rather, category label learning emerges from the online interaction between inhibitory processes and previously-learned items.

Dynamic Neural Field models are theoretically situated in Dynamic Systems theory, in which complex yet stable behavioural and cognitive systems emerge *ad hoc* from the interaction between components available at a given time (for example, the body, perceptual input, and the task environment) in the context of nested timescales of learning (for example, lifetime experience with categories and labels, exemplars and labels encountered earlier in the experiment, and the exemplar and label present on a given trial). Thus, these data add weight to the growing body of work demonstrating that cognition, behaviour and the environment are inextricably coupled and inseparable from their temporal context, contributing to our understanding of young children's categorisation, as well as to a new conception of developing cognition as an emergent dynamic system.

References

1. J. P. Spencer and G. Schöner, *Developmental Science* **6**, 392 (2003).
2. C. Houston-Price, K. Plunkett and P. Harris, *Journal of Child Language* **32**, 175 (2005).
3. W. V. O. Quine, *Word and object: An inquiry into the linguistic mechanisms of objective reference*. Cambridge: MIT Press. (1960).
4. L. B. Smith, in R. M. Golinkoff, K. Hirsh-Pasek, L. Bloom, L. B. Smith, A. L. Woodward, N. Akhtar, M. Tomasello and G. Hollich (Eds.), *Becoming a Word Learner: A Debate on Lexical Acquisition,* p. 55). New York: Oxford University Press (2000).
5. S. Carey and E. Bartlett, *Papers and Reports on Child Language Development* **15**, 17 (1978).
6. S. Carey, *Behavioral and Brain Sciences* **34**, 152 (2011).
7. E. Colunga and L. B. Smith, *Psychological Review* **112**, 347 (2005).
8. V. Gliozzi, J. Mayor, J. F. Hu, and K. Plunkett, *Cognitive Science* **33**, 709 (2009).
9. B. McMurray, J. S. Horst and L. K. Samuelson, L. K., *Psychological Review* **119**, 831 (2012).

10. L. K. Samuelson, L. B. Smith, L. K. Perry and J. P. Spencer, *PloS one* **6**, e28095 (2011).
11. V. R. Simmering, J. Triesch, G. O. Deak and J. P. Spencer, *Child Development Perspectives* **4**, 152 (2010).
12. K. E. Twomey and J. S. Horst, paper presented at the Society of Artificial Intelligence and the Simulation of Behaviour, University of York, UK (2011).
13. J. P. Spencer, M. S. C. Thomas, and J. L. McClelland, *Toward a Unified Theory of Development*: Oxford University Press New York (2009).
14. J. L. McClelland, M. M. Botvinick, D. C. Noelle, D. C. Plaut, T. T. Rogers, M. S. Seidenberg, et al., *Trends in Cognitive Sciences* **14**, 348 (2011).
15. B. A. Younger and L. B. Cohen, *Child Development* **54**, 858 (1983).
16. B. A. Younger and D. D. Fearing, *Infant Behavior and Development* **21**, 289 (1998)
17. P. C. Quinn, P. D. Eimas and S. L. Rosenkrantz, *Perception* **22**, 463 (1993).
18. T. T. Rogers and J. L. McClelland, *Semantic cognition: A parallel distributed processing approach*. Cambridge, MA: MIT Press (2004).
19. V. M. Sloutsky and A. V. Fisher, *Journal of Experimental Psychology: General* **133**, 166 (2004).
20. V. M. Sloutsky and A. V. Fisher, *Journal of Experimental Child Psychology* **111**, 65 (2012).
21. G. Diesendruck, L. Markson and P. Bloom, *Psychological Science* **14**, 164 (2003)
22. B. Landau, L. B. Smith and S. S. Jones, *Cognitive Development* **3**, 299 (1988).
23. L. K. Samuelson and J. S. Horst, *Infancy* **11**, 97 (2007).
24. L. B. Smith, S. S. Jones, B. Landau, L. Gershkoff-Stowe and L. Samuelson, *Psychological Science* **13**, 13 (2002).
25. C. Faubel and G. Schöner, *Neural Networks* **21**, 562 (2008).
26. J. S. Horst and L. K. Samuelson, *Infancy* **13**, 128 (2008).
27. M. Abecassis, M. D. Sera, A. Yonas and J. Schwade, *Journal of Experimental Child Psychology* **78**, 213 (2001).
28. A. L. Fulkerson and R. A. Haaf, *Infancy* **4**, 349 (2003).
29. E. Colunga and L. B. Smith, *Developmental Science* **11**, 195 (2008).
30. J. S. Horst and K. E. Twomey, *Infant and Child Development* (2012).
31. T. Klee and C. Harrison, paper presented at the Child Language Seminar, University of Hertfordshire (2001).
32. B. A. Younger and L. B. Cohen, *Child Development* **57,** 803 (1986).
33. K. Plunkett, J. F. Hu and L. B. Cohen, *Cognition* **106**, 665 (2008).
34. D. H. Rakison and L. B. Cohen, *Developmental Science* **2**, 423 (1999).
35. B. A. Younger, G. Hollich and S. D. Furrerm, *Infancy* **5**, 209 (2004).
36. V. R. Simmering, A. R. Schutte and J. P. Spencer, *Brain Research* **1202**, 68 (2008)
37. L. K. Samuelson, A. R. Schutte and J. S. Horst, *Cognition* **110**, 322 (2009).

THEORETICAL AND COMPUTATIONAL LIMITATIONS IN SIMULATING 3- TO 4-MONTH-OLD INFANTS' CATEGORIZATION PROCESSES

M. MERMILLOD

Laboratoire de Psychologie et Neurocognition, UMR CNRS 5105, Université Pierre Mendès-France
Grenoble, France

N. VERMEULEN

Université Catholique de Louvain and National Fund for Scientific Research, Psychology Department, B-1348
Louvain-la-Neuve, Belgium

G. KAMINSKY

Toulouse University, Laboratoire CLLE-LTC
Maison de la Recherche, France

E. GENTAZ

Laboratoire de Psychologie et Neurocognition, UMR CNRS 5105, Université Pierre Mendès-France
Grenoble, France

P. BONIN

Université de Bourgogne and CNRS
UMR 5022, France

It was shown that a simple connectionist autoencoder was able to simulate the asymmetric categorization effect reported by Quinn, Eimas and Rosenkrantz[3] in 3- to 4-month-old infants.[1,2] Moreover, other studies have reported that a careful control of the variance distribution and inclusion relationship of the inputs (meaning that a broader category subsumed a narrower visual category) is sufficient to reverse or even suppress the asymmetry.[4] In the current paper, we show that variance distribution and the inclusion relationship are not sufficient to produce the asymmetry. We show that (i) correlation information reported at the level of the 10 features previously encoded for connectionist simulations was actually very weak and (ii) if we increase these correlations among the different features, we completely suppress the asymmetry categorization effect, even with the same distribution and inclusion relationship that were assumed to produce the asymmetry.

1. Introduction

Science evolves by showing the limitations and constraints of theoretical models that are capable of explaining empirical results. The aim of this paper is to reveal the limitations and constraints of such a model that has been used for a decade to simulate categorization capacities in 3- to 4-month-old infants.[1,2,4,5]

At the empirical level, substantial progress in developmental psychology has made it possible to investigate visual categorization properties by means of experimental designs using familiarization/preference for novel paradigms. Based on these methodological frameworks, it has been shown that when presented with simple dot pattern stimuli, 3- to 4-month-old infants generalize looking times to novel instances from the familiar form category, and display visual preference for novel instances from novel form categories.[6] Further studies based on more complex realistic color images of basic-level animal categories, such as the cat and dog categories, have shown that 3- to 4-month-olds infants are able to form category representations including new exemplars from the same basic level but excluding exemplars from other basic level categories.[7] As suggested by other recent studies[8], this category formation process has been assumed to be based on the perceptual variability of the inputs.

A connectionist model by Mareschal and colleagues has attempted to simulate these perceptual categorization results observed in infants.[1,9] More precisely, the aim of this neural network was to reproduce the "asymmetric categorization effect".[3] The effect, based on visual basic level categories, was as follows: when infants were familiarized on the Cat category, they seemed to form a perceptual representation of cats that excludes dogs. On the other hand, when the infants were familiarized on the category Dog, they were not able to categorize a novel cat as coming from a novel category. That is, the Cat and Dog categories have asymmetric exclusivity: Cat excludes dogs, but Dog does not exclude cats.

On the basis of connectionist modeling, previous researchers[1,2,4,5] have proposed a connectionist autoencoder which makes it possible to replicate the effect showed on 3- to 4-month-olds. The analogy between the error produced by the neural network and infant's looking time was as follows. During fixation, the infant is believed to be engaged in a process of interactive representation construction.[10] The infant ceases to examine a stimulus once he or she has developed a reliable internal representation of that stimulus. Similarly, learning in an autoencoder consists of an iterative process of representation construction. The network tries to adjust its internal representation of the input (i.e., by adjusting the network weights) until it encodes enough information about the input stimulus to be able to reconstruct that stimulus in the output. The central assumption of this account is that stimuli that produce a high level of initial output errors (that are poorly autoencoded) will take longer to encode (reduce the output error) than stimuli that produce low initial output error rates. This is because the network will require more iterations to adjust the internal representations appropriately. Hence, the output error produced when a novel stimulus is presented to the network is taken as an equivalent to the time the infant will spend looking at the stimulus. The model was successful in capturing the category-based looking time behaviors of 3- to 4-month-olds, including the subtle asymmetric exclusivity in the extensions of the categories tested, such as Cat and Dog. It has been suggested[1,2,3,4] that the exclusivity difference might reflect an asymmetric relation in the distribution of feature values used to characterize the two sets of images. Indeed, they found that, in the image set used by Quinn et al.[3], the large majority of cat feature values were subsumed within the distribution of dog feature values. In addition, the majority of dog feature values did not fall within the distribution of cat feature values. Thus, at the level of individual features, most cats were plausible dogs, but most dogs were not plausible cats. This asymmetry effect could explain the observed behaviors if infants are assumed to acquire their looking time behaviors solely on the basis of the features encountered during familiarization, as was the case with the networks.

Interestingly, Furrer and Younger[11] reported that the asymmetry categorization effect completely disappears for 10-month-old infants. Research was therefore conducted in the field of developmental psychology to investigate the shift in the asymmetry effect in the behavior of 3- to 4-month-old infants to that of 10-month-old infants.[11] In addition, a variety of papers have shown that adults[12], as well as infants as young as 10 months of age[13], are able to use information about correlations among the different features of the stimuli to produce correct categorizations of visual and natural objects. Conversely, 4-month-old infants do not seem to be able to use correlation information to perform categorization tasks[13]. Therefore, investigating the response of the theoretical model proposed in the literature to correlated attributes is, on the one hand, important in the light of the fact that this information is broadly used by humans to perform categorization[12,13] and, on the other, is crucial when we consider that most natural categories have features that naturally covary. This idea is supported by a number of papers published in the biological sciences which suggest that the size of different exemplars of any given species remains constant between scales[14], thus leading strong correlations among the different features of the exemplars. Moreover, this characteristic of living categories is an old and well-known property in the life sciences and is often used for the characterization of different species[15] or studies of the maturation of living organisms.[16] This property means, for example, that even though it has been shown that category "dog" is very variable[1,2], there will be significant correlations among the different features of any given dog: a big dog will have a big head, a big body and big legs. An absence of any correlation among features would mean that we could not infer from the size of a poodle's head whether its legs or body will be smaller or bigger than those of a great dane.

Our prediction here is that the ability (or inability) of the theoretical model to simulate the asymmetry is highly dependent on the absence (or presence) of correlated attributes. To test this, we ran 3 connectionist simulations. Simulation 1 was designed to reproduce the asymmetry of the original data. However, we also addressed the possibility that the method used to encode the stimuli (measurement of features of 2D pictures) could produce weak correlation among features. Although, of course, this is only an assumption since we no longer have access to the real 3D animals, we will show that the data used for the simulation had a low correlational structure. The aim of Simulation 2 was to show that the original model proposed by French, Mareschal, Mermillod and Quinn[4], is able to differentiate between two categories if we artificially introduce correlational information in the neural network. We have to remember here that correlation information was not tested in the original paper[4] which focused primarily on the effect of perceptual variability on categorization asymmetry. However, at a computational level, it has been shown that neural networks are able to use both perceptual variability and also information about correlations among different dimensions for the successful categorization of complex natural stimuli.[17] Our assumption here therefore is that, like human adults[12] or 10 month-old infants[13], neural networks are able to use correlation information independently of the perceptual variability of the features to perform the categorization task in a reliable manner. We will show on the basis of new computational data that the asymmetric categorization effect completely disappears in the connectionist network when exposed to a set of correlated attributes. These effects were obtained despite variance distributions that were previously assumed to generate the asymmetry. The aim of Simulation 3 was to show that the correlations among features must be completely eliminated in this model to simulate the expected asymmetry observed in 3- to 4-month-old infants.

2. Simulation 1. Reproduction of the asymmetric categorization effect

The aim of this first simulation was to reproduce the asymmetry categorization effect and provide measurements of the correlation between the ten features in the original data. The stimuli, neural networks and procedure were strictly identical to French et al.[4] As in these previous simulations, correlation information was controlled to be very weak across these features.

2.1. *Stimuli*

The data consists of the feature measurements from 18 photographs of dogs and 18 photographs of cats[4]. The photographs were taken from Siegal[18] and Schuler[19] in order to obtain dogs with low perceptual variance and cats with high variance (as shown on figure 1). It should be remembered here that we were simulating the most recent results reported by French et al.[4] In this paper, the variability of the cat and dog categories was reversed compared to Quinn et al.[3], with the result that the dog category was narrower and subsumed by the broader cat category. However, this does not change the results of the computational simulations since we manipulated matrices of data with different variance distributions and correlations and the category name – "cat" or "dog" – is merely a label that could be substituted for any other. As in the original simulations, the networks were trained on the same ten measurements of the stimuli used to familiarize the infants (i.e., horizontal extent, vertical extent, leg length, head length, head width, eye separation, ear separation, nose length, nose width).

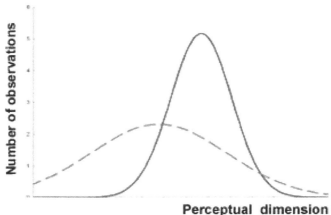

Fig. 1. Example of one of the feature distributions (among the ten features) for exemplars in Simulation 1. The cat distributions (dashed line) subsume the dog distributions (solid line).

2.2. Neural network procedure

We used exactly the same standard 10-8-10 back-propagation autoencoder network used by French et al.[4] to model infant categorization, meaning that we had 10 input and output units (the measurements of the ten features) and 8 hidden units. Training parameters were identical to the original simulation. The vectors used for this simulation were normalized between 0 and 1, feature by feature, across all of the 36 stimuli (i.e., 18 dogs and 18 cats). The networks were trained in six batches of two patterns until all outputs were within 0.2 of their target values or for a maximum of 250 weight updates. Each batch of 2 patterns was trained fully before presenting the next batch of two. This procedure was intended to mimic the paired presentation technique used with infants. The error, as in the original simulations[1,9], was the maximum MSE produced by the activation of the output nodes compared to the desired output. The results were averaged over 50 neural networks with different initial random weights. Each network was first trained on twelve randomly selected exemplars selected from either the set of cat images or, alternatively, the set of dog images. Once a network had learned to autoencode these images to criteria, it was tested on 6 novel images from each of the two categories. The amount of output error produced in response to the novel images is a measure of the novelty of the images and is assumed to reflect the infant's looking time.

2.3. Results

Figure 2 shows the average network error in response to the novel test stimuli for both possible training regimes. It should be recalled here that the simulation was conducted on the stimuli used by French et al.[4] in which the variance and inclusion relationships were reversed, with the result that the broader cat category subsumed the narrower dog category. As in the original paper, we conducted an analysis of variance (ANOVA) on neural network error rate as a dependent variable, and Training Category (Cat or Dog category) x Test Category (Cat or Dog category) as a within-subjects independent variable. ANOVA revealed a significant interaction between the type of training category and the test category, $F(1, 98) = 177.5$, $p<0.001$, $\eta 2 =.56$). Moreover, pairwise comparisons showed that, after training on the broad Cat category, there was no significant difference between the two test categories, $F(1, 98) = 2.03$, $p>0.16$. Thus, it appears that they have formed a category of cat that does not differentiate dogs from cats, as in French et al.[4]

Networks trained on dogs produced a greater amount of error (meaning a greater amount of looking time for infants) to the novel cat than the novel dog category, $F(1, 98) = 411.2$, $p<0.001$. Thus, it appears that they have formed a category of Dog that excludes novel cats, which is reproducing the original asymmetry reported in French et al.[4] at a connectionist and behavioral level.

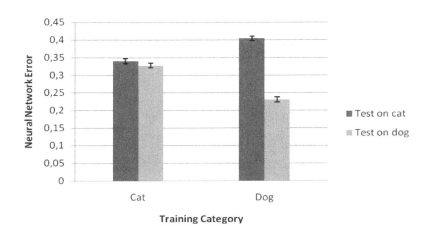

Fig. 2. Average network error produced in response novel dogs or cats following training with either the Cat or Dog category.'

2.4. *Correlation analysis*

The aim of this part was to analyze the effect of correlation among the different features on categorization performance at a computational level, in relationship with variability distribution of these features. Therefore, we measured the correlation matrix among all features of this original Simulation 1. As shown in table 1a and 1b, the correlational structure of the Cat and Dog categories actually seems to be very weak.

Table 1a. Auto-correlation of the category Dog used in Simulation 1

	Head length	Head width	Eye separation	Ear separation	Ear length	Nose length	Nose width	Leg length	Vertical extent	Horizontal extent
Head length	1	0.673	0.056	0.248	-0.01	0.293	0.24	0.038	0.528	-0.37
Head width	0.673	1	0.383	0.4	0.028	-0.31	0.05	-0.46	0.153	-0.14
Eye separation	0.056	0.383	1	1.00E-16	0.354	-0.39	-0.15	-0.61	-0.1	0.273
Ear separation	0.248	0.4	1.00E-16	1	0.186	0.166	0.37	-0.14	-0.24	0.134
Ear length	-0.01	0.028	0.354	0.186	1	0.019	-0.18	-0.26	-0.48	0.158
Nose length	0.293	-0.31	-0.39	0.166	0.019	1	0.539	0.552	0.285	-0.07
Nose width	0.24	0.05	-0.15	0.37	-0.18	0.539	1	0.468	0.296	-0.13
Leg length	0.038	-0.46	-0.61	-0.14	-0.26	0.552	0.468	1	0.229	-0.21
Vertical extent	0.528	0.153	-0.1	-0.24	-0.48	0.285	0.296	0.229	1	-0.21
Horizontal extent	-0.37	-0.14	0.273	0.134	0.158	-0.07	-0.13	-0.21	-0.21	1

Table 1b. Auto-correlation of the category Cat used in Simulation 1

	Head length	Head width	Eye separation	Ear separation	Ear length	Nose length	Nose width	Leg length	Vertical extent	Horizontal extent
Head length	1	0.669	0.498	0.456	-0.07	0.548	0.434	-0.01	0.588	0.736
Head width	0.669	1	0.756	0.758	0.165	0.181	0.639	0.345	0.847	0.319
Eye separation	0.498	0.756	1	0.785	0.551	0.367	0.748	0.189	0.542	0.086
Ear separation	0.456	0.758	0.785	1	0.5	0.2	0.655	0.28	0.564	0.017
Ear length	-0.07	0.165	0.551	0.5	1	0.171	0.431	0.234	0.158	-0.37
Nose length	0.548	0.181	0.367	0.2	0.171	1	0.507	-0.27	-0.04	0.209
Nose width	0.434	0.639	0.748	0.655	0.431	0.507	1	0.165	0.415	-0.11
Leg length	-0.01	0.345	0.189	0.28	0.234	-0.27	0.165	1	0.27	0.012
Vertical extent	0.588	0.847	0.542	0.564	0.158	-0.04	0.415	0.27	1	0.412
Horizontal extent	0.736	0.319	0.086	0.017	-0.37	0.209	-0.11	0.012	0.412	1

2.5. *Discussion*

These results show that simple autoencoder networks are able to form perceptual categories of Cat and Dog that reflect those of 3 to 4-month-olds when familiarized and tested with the same images as the infants. However, the asymmetry was based on the original measurement data which had a weak correlational structure. Similar correlational structures were obtained on the first round of connectionist simulations[1,2,9] based on the original data[3]. These original data do not allow us to test the response of the connectionist networks when exposed to correlated data. Hence, we artificially modified the original data in order to obtain a new pattern of correlated data while keeping the same variance distribution that had induced the asymmetry in the original simulations. Given that most natural categories have correlated features and since our cognitive system is able to use this information to distinguish between different natural categories[12], we artificially increased the correlation information in the original data. Therefore in the current theoretical paper, we used artificial categories to explore the computational properties of the model with regard to this type of correlated attribute.

3. Simulation 2. The asymmetric categorization effect with correlated exemplars

In this simulation, we artificially modified the pattern of data (i.e. the feature values of the original stimuli used in Simulation 1 as input/output for neural network simulations) in order to increase correlations among features within each category. We assumed that the introduction of correlation information might allow the neural network to differentiate the original categories despite their variance and inclusion relationship.

3.1. *Stimuli*

The pattern was generated by taking the mean of each category for each feature, and then multiplying these means by a discrete value in order to obtain two perfectly auto-correlated categories. We then added random noise to the two category exemplars in order to obtain two statistically variable set of features. The crucial point is that these data retained the same variance distribution and inclusion relationship of each category that was assumed to generate the asymmetry. The result at the level of the inputs of the system is that correlations were artificially controlled to ensure that they were greater for each category than in simulation 1 (table 2a and 2b).

Table 2a. Auto-correlation of the category Dog used in Simulation 2

	Head length	Head width	Eye separation	Ear separation	Ear length	Nose length	Nose width	Leg length	Vertical extent	Horizontal extent
Head length	1	0.88	0.47	0.79	0.66	0.66	0.29	0.87	0.88	0.93
Head width	0.88	1	0.46	0.83	0.66	0.73	0.4	0.91	0.87	0.9
Eye separation	0.47	0.46	1	6.50E-01	0.6	0.42	0.5	0.52	0.64	0.6
Ear separation	0.79	0.83	6.50E-01	1	0.61	0.72	0.53	0.86	0.92	0.92
Ear length	0.66	0.66	0.6	0.61	1	0.49	0.43	0.74	0.71	0.74
Nose length	0.66	0.73	0.42	0.72	0.49	1	0.38	0.72	0.68	0.73
Nose width	0.29	0.4	0.5	0.53	0.43	0.38	1	0.41	0.39	0.43
Leg length	0.87	0.91	0.52	0.86	0.74	0.72	0.41	1	0.93	0.95
Vertical extent	0.88	0.87	0.64	0.92	0.71	0.68	0.39	0.93	1	0.95
Horizontal extent	0.93	0.9	0.6	0.92	0.74	0.73	0.43	0.95	0.95	1

Table 2b. Auto-correlation of the category Cat used in Simulation 2

	Head length	Head width	Eye separation	Ear separation	Ear length	Nose length	Nose width	Leg length	Vertical extent	Horizontal extent
Head length	1	0.97	0.8	0.94	0.94	0.84	0.58	0.98	0.97	0.97
Head width	0.97	1	0.82	0.95	0.94	0.86	0.62	0.96	0.98	0.98
Eye separation	0.8	0.82	1	7.90E-01	0.87	0.8	0.52	0.83	0.86	0.83
Ear separation	0.94	0.95	7.90E-01	1	0.92	0.89	0.59	0.97	0.97	0.97
Ear length	0.94	0.94	0.87	0.92	1	0.88	0.62	0.95	0.96	0.95
Nose length	0.84	0.86	0.8	0.89	0.88	1	0.58	0.85	0.88	0.87
Nose width	0.58	0.62	0.52	0.59	0.62	0.58	1	0.59	0.62	0.63
Leg length	0.98	0.96	0.83	0.97	0.95	0.85	0.59	1	0.99	0.99
Vertical extent	0.97	0.98	0.86	0.97	0.96	0.88	0.62	0.99	1	0.99
Horizontal extent	0.97	0.98	0.83	0.97	0.95	0.87	0.63	0.99	0.99	1

3.2. *Network*

The simulations were strictly identical to simulation 1 but with the new pattern of data presented in Table 2a and 2b.

3.3. *Results*

As in Simulation 1, we observed a significant interaction between the type of training category and the test category effect, $F(1, 98) = 164.19$, $p<0.001$, η^2 =.59. The model was able to produce a significant increase in errors for dogs after training on the broad Cat category, $F(1, 98) = 74.34$, $p<0.001$. However, after being trained on the narrow Dog category, the model produced a greater amount of error on the novel cat than the novel dog category, $F(1, 98) = 90.23$, $p<0.001$. As expected, the correlation information provided by this new pattern of data was sufficient to suppress the asymmetry despite the variance distribution and inclusion relationship (figure 3).

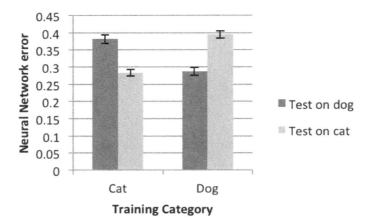

Fig. 3. Average network error produced in response to novel dogs or cats following training with either the Cat or Dog category.

3.4. *Discussion*

No asymmetry occurred despite the variance and inclusion relationship in the data. The back-propagation auto-encoder seems to be able to use the correlational structure of the two categories to identify the exemplars as coming from different categories. Therefore, it seems that the original model revealed an asymmetric categorization effect because there were poor correlations in the input data. This could be due to the fact that the Cat and Dog categories do not actually have strong correlations among their features in the real world or that the features taken from the stimuli do not account for such existing correlations. The latter hypothesis might be due to the simple fact that it could be difficult to encode 3-dimensional feature measurements on the basis of their 2-dimensional image representation. The features extracted in 2-D images depend on the projection of the 3-D objects and there is therefore a possible loss of correlation information among features of similar exemplars belonging to the same category. To test this assumption, we used the neural network on a pattern of uncorrelated data. We assumed that this pattern is able to produce asymmetries in a similar way to the original data.

4.   Simulation 3. The asymmetric categorization effect after removal of the category correlations

The purpose of this simulation was to control that the presence of a low correlational structure within each category was a crucial factor allowing the asymmetry. Therefore, we assumed that completely removing the correlational structure would not prevent the network from reproducing the original asymmetry.

4.1. *Stimuli*

In order to control the effect of correlation in the auto-encoder, we modified the pattern of data used by French et al.[4] We computed the average value for each feature in each category and then multiplied this mean by a different random number for each stimulus and each feature in order to obtain a randomly uncorrelated pattern of data. As in the previous simulations, we kept a comparable variance distribution and inclusion relationship for each category required in order to obtain the asymmetric categorization effect. The correlations for each category were removed to provide two patterns of uncorrelated data (table 3a and 3b).

Table 3a. Auto-correlation of the category Dog used in Simulation 3

	Head length	Head width	Eye separation	Ear separation	Ear length	Nose length	Nose width	Leg length	Vertical extent	Horizontal extent
Head length	1	0.055	0.142	0.591	-0.47	0.353	-0.11	0.031	0.095	0.096
Head width	0.055	1	-0.25	-0.19	0.148	0.02	0.337	-0.09	0.103	0.034
Eye separation	0.142	-0.25	1	-1.00E-01	-0.26	0.034	-0.01	0.143	-0.37	0.205
Ear separation	0.591	-0.19	-1.00E-01	1	-0.2	0.092	0.042	0.075	0.102	-0.07
Ear length	-0.47	0.148	-0.26	-0.2	1	0.024	-0.06	-0.11	-0.06	-0.19
Nose length	0.353	0.02	0.034	0.092	0.024	1	-0.26	-0.17	-0.05	0.31
Nose width	-0.11	0.337	-0.01	0.042	-0.06	-0.26	1	0.152	-0.25	0.227
Leg length	0.031	-0.09	0.143	0.075	-0.11	-0.17	0.152	1	0.292	0.043
Vertical extent	0.095	0.103	-0.37	0.102	-0.06	-0.05	-0.25	0.292	1	-0.22
Horizontal extent	0.096	0.034	0.205	-0.07	-0.19	0.31	0.227	0.043	-0.22	1

Table 3b. Auto-correlation of the category Cat used in Simulation 3

	Head length	Head width	Eye separation	Ear separation	Ear length	Nose length	Nose width	Leg length	Vertical extent	Horizontal extent
Head length	1	0.03	-0.03	-0.04	0.163	0.404	-0.25	0.031	-0.42	0.301
Head width	0.03	1	-0.16	-0.11	0.199	0.176	0.04	-0.21	0.279	0.163
Eye separation	-0.03	-0.16	1	6.50E-02	-0.19	0.252	0.265	-0.04	-0.01	-0.08
Ear separation	-0.04	-0.11	6.50E-02	1	-0.02	-0.01	0.069	0.459	-0.38	0.066
Ear length	0.163	0.199	-0.19	-0.02	1	-0.4	-0.49	-0.44	-0.11	0.195
Nose length	0.404	0.176	0.252	-0.01	-0.4	1	-0.06	0.162	0.148	0.296
Nose width	-0.25	0.04	0.265	0.069	-0.49	-0.06	1	0.22	0.055	-0.32
Leg length	0.031	-0.21	-0.04	0.459	-0.44	0.162	0.22	1	-0.03	-0.24
Vertical extent	-0.42	0.279	-0.01	-0.38	-0.11	0.148	0.055	-0.03	1	-0.22
Horizontal extent	0.301	0.163	-0.08	0.066	0.195	0.296	-0.32	-0.24	-0.22	1

4.2. *Neural network procedure*

This simulation was run using exactly the same networks and parameters as simulation 1 and 2.

4.3. *Results*

As in the original data, the variance distribution and inclusion relationship produced a reliable asymmetry (figure 4). The two-way interaction between the type of training category and the test category was significant, $F(1, 98) = 584.5$, $p<0.001$, $\eta^2 = .16$. After training on the broad Cat category, planned comparisons of the means revealed that there was no significant difference between the two test categories, $F(1, 98) = .473$, $p>0.493$. Networks trained on Dog category produced a greater amount of error on the novel cat than the novel dog category, $F(1, 98) = 1122.5$, $p<0.001$. Thus, it appears that they formed a category of Dog that excludes novel cats, thus reproducing the original asymmetry.

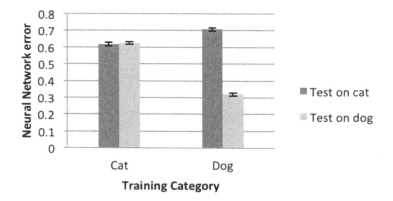

Fig. 4. Average network error produced in response to novel dogs or cats following training with either the Cat or Dog category.

4.4. *Discussion*

The asymmetry was perfectly reproduced, thus suggesting that correlation is a major factor determining the presence or absence of the asymmetry. In simulation 3, as opposed to simulation 2 in which structures were highly correlated, uncorrelated structures allowed the asymmetry categorization effect to emerge in the categorization model applicable to 3- to 4-month-old infants.

5. Conclusion

These data clearly show that the manipulation of the distribution and inclusion relationship is not in itself sufficient to determine the asymmetric categorization effect reported by Quinn et al.[3] The model proposed by Mareschal and French[5] which at the moment is regarded as one of the standard representation of categorization in 3- to 4-month-olds may in fact only be a good approximation to this age group's performance in the specific case of uncorrelated stimuli. Feature correlation as a factor was not considered in the original model, but new simulations with correlated vs. uncorrelated feature sets suggest that it has a massive impact on model performance. As the sensitivity to feature correlation is further one of the main factors to change during infancy this led to suggest a research agenda for further studies investigating the role of feature correlation in symmetric vs. asymmetric categorization across development. As demonstrated in simulation 2, the asymmetric effect completely disappears when correlated attributes are present within each category. These results show that the network's original asymmetry[1,4,9,20] was produced by the weak correlational structure of the visual categories (Simulation 1). An hypothesis that can be proposed here is that this lower correlation among features might be due to the 2-Dimensional projections of the 3-Dimensional features constituting each category.

Another way to test these hypotheses at the empirical level is to use stimuli similar to Younger and Cohen[13], namely artificial animals comprising a set of perceptually controlled features (legs, body, neck, head, etc.) Regarding the visual categorization properties at a behavioral level, 3- to 4-month-old infants do not seem to be sensitive yet to correlation information.[13,21] Therefore, this behavioral characteristic makes them sensitive only to the variance distribution and inclusion relationship, irrespective of the correlational structure of the categories, thus allowing - de facto - the asymmetry to appear. However, these data raise several questions in connection with 10-month-old infants.[13,22] In particular, if 10-month-olds are able to use the correlational structure of the categories to differentiate between the narrow and included categories, it should be possible to suppress the asymmetry by controlling the correlations within each category. This hypothesis supports and extends empirical evidence showing that the asymmetry categorization effect was reproduced for 4-month-old infants but disappeared in some cases in 10-month-old infants.[11] However, the current literature lacks a theoretical explanation for the shift from asymmetric categorization in 3- to 4-month-old to symmetric categorization in 10-month-old infants. Furrer and Younger[11] assumed that the disappearance of the asymmetry was due to the increased speed or efficiency of processing or to prior exposure to pets.[23,24] Here we provide a theoretical model, based on neural computation, which suggests that the theoretical underpinning for the shift from asymmetric categorization to symmetric categorization is related to the use of correlation information by 10-month-old infants. Moreover, we show that the theoretical model proposed by French et al.[4] is able to account for 10-month-old infants' categorization capacities (but finds it more difficult to account for those displayed by 3- to 4-month-old infants). Therefore, and contrary to the prevailing approach in developmental psychology, subsequent research into neural computation will have to focus more on 3- to 4-month-old infants than 10-month-old infants. Ways of simulating 3- to 4-month-old infants' categorization asymmetry on categories possessing correlated features still need to be developed.

Our current connectionist data propose a theoretical framework specifying how humans are able to suppress this type of asymmetry in categorization by shifting from variability-based categorization to correlation-based categorization in the first months of life. More precisely, the model suggests that the asymmetry should disappear in 10-month-old infants, even without prior exposure to pets[23,24], if both categories have a correlational structure. In other words, if 10-month-olds only use the correlation information to differentiate the two categories, the model suggests that the asymmetry observed in 3- to 4-month-old infants should remain when a pattern of uncorrelated data is used with older infants, even in the case of complex stimuli such as the cat and dog categories. As far as 3- to 4-month-old infants are concerned, we should observe an asymmetric categorization effect in both conditions.

Therefore, based on further behavioral experiments in 3- to 4-month-old and 10-month-old infants, it will probably be necessary to modify our connectionist architecture to account for both empirical conditions. One possible way to explain such behavioral data is that, unlike in the case of variance distribution, biological or artificial neural systems might need neural maturation in order to process correlation information. The mainstream of research in visual categorization based on prototype models[25] to exemplar models[26,27,28] or even connectionist models[4,17,29,30,31,32] has mainly focused on variance distribution. However, a number of papers have shown that, at the empirical level, correlation information is important for stimulus categorization.[12,24] For our purposes, we can assume that the variance distribution of the stimuli is necessary to generate an internal representation of the categories (see French et al.[4], Mermillod, Vermeulen, Lundqvist and Nidenthal[17]] for clear examples of this process) but that correlation information may subsequently distort this internal representation and thereby improve categorization efficiency. This hypothesis will need to be addressed in future studies in the cognitive sciences if we are to gain a better understanding of the combined effect of these two variables in categorization processes.

Acknowledgments

This work was supported by the French CNRS (UMR 6024), the InstitutUniversitaire de France, a grant from the French National Research Agency (ANR Grant BLAN06-2_145908, ANR Grant BLAN08-1_353820), and a Clermont Université PHRC program to MM.

References

1. D. Mareschal, R. M. French and P. C. Quinn, *Developmental Psychology* **36**, 635 (2000).
2. D. Mareschal, P. C. Quinn and R. M. French, *Cognitive Science* **26**, 377 (2002).
3. P. C. Quinn, P. D. Eimas and S. L. Rosenkrantz, *Perception* **22**, 463 (1993).
4. R. M. French, D. Mareschal, M. Mermillod and P. C. Quinn, Journal of Experimental Psychology : General **133**, 382 (2004).
5. D. Mareschal and R. M. French, *Infancy* **1**, 59 (2000).
6. P. C. Quinn, *Cognition* **27**, 145 (1987).
7. P. D. Eimas and P. C. Quinn, *Child Development* **65**, 903 (1994).
8. H. Kloos and V. M. Sloutsky, *Journal of Experimental Psychology: General* **137**, 52 (2008).
9. D. Mareschal and R. M. French, In M. G. Shafto & P. Langley (Eds.), *Proceedings of the 19th annual conference of the Cognitive Science Society,* 484. Mahwah, NJ: Erlbaum. (1997).
10. E. N. Sokolov, *Perception and the conditioned reflex.* Hillsdale, NJ: Erlbaum (1963).
11. S. D. Furrer and B. A. Younger, *Developmental Science* **8**, 544 (2005).
12. J. R. Anderson and J. M. Fincham, *Journal of experimental psychology: Learning, Memory and Cognition* **22**, 259 (1996).
13. B. Younger and L. B. Cohen, *Child Development* **57**, 803 (1986).
14. B. B. Mandelbrot, *The Fractal Geometry of Nature.* New York: Freeman (1977).
15. P. Cavallini, *Annales Zoologici Fennici* **32**, 421 (1995).
16. P. C. Hughes and J. M. Tanner, *Journal of Anatomy* **106**, 349 (1970).
17. M. Mermillod, N. Vermeulen, D. Lundqvist and P. M. Niedenthal, *Cognition* **110**, 346 (2009).
18. M. Siegal (Ed.), *Simon & Schuster's guide to cats.* New York: Simon & Schuster (1983)
19. E. M. Schuler (Ed.), *Simon & Schuster's guide to dogs.* New York: Simon & Schuster (1980).
20. R. M. French, M Mermillod, P. C. Quinn and D. Mareschal, In K. Stenning & J. Moore (Eds.), *Proceedings of the 23rd annual conference of the Cognitive Science Society,* 307. Mahwah, NJ: Erlbaum (2001).
21. B. Younger and L. B. Cohen, *Child Development* **54**, 858 (1983).

22. B. Younger, *Child Development* **56**, 1574 (1985).
23. K. B. Hurley, K. A. Kovack-Lesh and L. M. Oakes, *Infant Behavior and Development* **33**, 619 (2010).
24. K. A. Kovack-Lesh, J. H. Horst and L. M. Oakes, *Infancy* **13**, 285 (2008).
25. E. Rosch, In: E. Rosch and B.B. Lloyd (Eds). *Cognition and Categorization*. Hillsdale, NJ: Lawrence Erlbaum (1978).
26. R. M. Nosofsky, *Journal of experimental psychology: General* **115**, 39 (1986).
27. J. K. Kruschke, *Psychological review* **99**, 22 (1992).
28. N Vermeulen, M. Mermillod, J. Godefroid and O. Corneille, *Cognition* **112**, 467 (2009).
29. M. Mermillod, P. Vuilleumier, C. Peyrin, D. Alleysson and C. Marendaz, *Connection Science* **21**, 75 (2009).
30. M. Mermillod, P. Bonin, L. Mondillon, D. Alleysson and N. Vermeulen, *Neurocomputing* **73**, 2522 (2010).
31. G. Kaminski, D. Méary, M. Mermillod and E. Gentaz, *Attention, Perception and Psychophysics* **73**, 1344 (2011).
32. G. Kaminski, D. Méary, M. Mermillod and E. Gentaz, *Perception* **39**, 807 (2010).

REINFORCEMENT-MODULATED SELF-ORGANIZATION IN INFANT MOTOR SPEECH LEARNING

ANNE S. WARLAUMONT

*Cognitive and Information Sciences, School of Social Sciences, Humanities, and Arts,
University of California, Merced, 5200 North Lake Rd.
Merced, CA 95343, USA
E-mail: awarlaumont2@ucmerced.edu
annewarlaumont.org*

Neural network models of early motor speech vocal learning are reviewed, with a focus on those models that utilize reinforcement to modulate what would otherwise be self-organized learning. It is argued that such a mechanism likely plays a role in bringing about the changes observed in prespeech vocalizations produced by human infants. Such models complement well the already popular purely self-organized learning models that focus on effects of exposure to sounds from the ambient language.

Keywords: Reinforcement, Self-organization, Speech-language development, Infancy, Babbling, Self-organizing map, Spiking neural networks

1. Introduction

Connectionist models can for the most part be grouped into two broad categories, those that learn under supervision, meaning that they receive explicit feedback about what their correct outputs should be, and those that learn in an unsupervised or self-organized fashion, meaning that they are given no information about what their outputs ought to be. The focus of the present paper is on a third way of providing feedback to connectionist models, reinforcement, and the valuable role that it may play in motor speech learning. I will argue that reinforcement is particularly useful in this domain because it relaxes some unrealistic assumptions made by supervised approaches to motor learning while at the same time providing structure that is lacking in motor exploration compared to perceptual learning. Reinforcement can be incorporated through a very simple modification to standard unsupervised learning approaches: increasing the learning rate when reinforcement is received. After reviewing some supervised and unsupervised approaches in the domain of early motor speech development, I will present two recent examples of the reinforcement-modulated learning approach from that same domain.

2. Background

There are many ways in which infant vocalizations change, apparently as a consequence of learning, across the first year or two of life, as the repertoire expands in pitch, amplitude, and vocal quality, comes to include syllabi-cally patterned vocalization, and eventually comes to include specific word forms[1]. A number of modelers interested in providing neural network ac-counts for some of these changes have focused their efforts on showing how unsupervised learning might contribute to the development of vocal imi-tation abilities. For example, one typical approach is to model the infant as having two layers of neurons, a sensory layer and a motor layer, con-nected via self-organizing Hebbian connections. The sensory and motor lay-ers themselves can be treated as self-organizing maps or can be given static connections to sensory input feature vectors and motor actuators. Training typically proceeds by producing random motor outputs, running those out-puts through a vocalization synthesizer, hearing the resulting sounds, then allowing self-organized learning to take place. Such models have demon-strated the ability to learn to imitate sounds[2-5] so that, for example, when a model hears the vowel /a/, activation propagates from the sensory repre-sentation of that sound to the motor system and the model then produces a similar sound.

However, development of the ability to imitate sounds that are already in the infant's repertoire is only one feature of prespeech vocal development. Other work has addressed how infants might come to learn to produce par-ticular sounds when a representation of a specific speech sound is activated. One approach that has been taken is to provide a neural network with input representing a target speech sound and then to have the network produce a motor output in response, training the network using supervision, where the correct motor commands to produce the input sound are given directly to the model[6]. A problem with this supervised approach is that it assumes that the infant somehow has access to explicit information regarding the correct motor commands that go with a particular speech sound. This seems very unlikely to be the situation faced by human infants. In fact, that in-formation is precisely the knowledge that the model is trying to learn. So although the model performs well, and proves that it is possible to encode the sensory-motor relationship for speech sounds in neural network weights, its learning is not as realistic as it should be.

A different approach to learning associations between speech sound representations and motor commands is to have a model produce random vocal babble and when it happens by chance to produce a sound from a particular speech sound category allow the network to learn an association between the randomly generated motor commands and the speech sound category[7]. This approach makes more realistic assumptions about the knowledge infants possess prior to learning. The approach could be classified as an unsupervised model (ignoring the issue of learning which sounds are speech sound targets in the first place).

Another important feature of early vocal development is that infants often produce sounds seemingly rather spontaneously, not in direct imitation of an adult nor apparently targeting a particular adult speech sound. And these spontaneously produced sounds change as the child gets older, coming to display more of the characteristics of adult vocalization. There are at least two ways in which such changes in spontaneously produced vocalizations can be effected. One way is for infant vocalizations to be influenced by targets, as in the DIVA model. By having a model receive auditory input from another speaker and by having this affect vocalization production (so that spontaneous vocalizations are now not purely random, but even during learning are affected by external input), a model can end up producing sounds that resemble that external input's[8,3]. This seems plausible, and would also be classified as a type of unsupervised learning. Figure 1 illustrates how some of the mechanisms reviewed here relate to each other and to those described below.

3. Reinforcement-modulated vocal learning with a self-organizing map

Another way to generate meaningful changes in spontaneously produced vocalizations is for the model's own vocalizations to be differentially reinforced and for that reinforcement to guide learning. One simple example of this can be found in a study in which my colleagues and I used a self-organized map (SOM) to control muscle activations in a vocal-tract-based vocalization synthesizer[9]. Typically, self-organizing maps are used in perceptual tasks. A single layer of neurons is connected to a vector of inputs. The neural network is exposed to inputs from a training corpus (in models of speech perception, these could be acoustic representations of various vowels in a language, for example) and with each input, adjusts its neurons' receptive fields (i.e. weighted connections between inputs and neurons) in a self-organizing fashion. SOMs in these cases capitalize on the structure

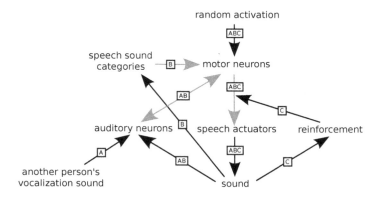

Fig. 1. Summary of key operations of some previous neural network models of early motor speech development. The gray arrows represent neural connections that can undergo self-organized learning. Black connections typically have been held constant over the course of learning in these models. Connections marked "A" are incorporated in models that self-organize auditory-motor connections on the basis of babbling and adult input. They learn to imitate adult sounds and their spontaneous vocalizations can come to resemble heard adult sounds. Connections marked "B" are those that incorporated in models that learn to associate certain motor patterns with speech sound categories (so far models have tended to assume that those categories' acoustic features are known beforehand). Connections marked "C" are incorporated in the models discussed below that learn through purely motor babble with learning modulated by reinforcement that is provided on the basis of a vocalization's acoustics.

inherent in the input, i.e. the kinds of vectors out of all possible vectors that are represented and their similarity to one another. When using a self-organizing map for motor learning, when neurons are not connected to perceptual inputs but rather to motor outputs, such as muscles, this inherent structure may not be present. In particular, models of early motor speech development such as many of those mentioned above have tended to have training trials with random motor "babble" at their core. Muscles are randomly activated and the resulting sounds are observed. In models where the goal is to form sensory-motor associations, this is not a problem, but if the goal is, without any auditory examples, to change the motor primitives themselves and in turn change the types of sounds that are produced during babble, there is a problem in that random motor output provides no interesting structure to self-organize around. Our study addressed this issue by introducing reinforcement when the model produced desirable sounds and by having learning occur only when reinforcement had been given. That is, learning was modulated by reinforcement.

Like the other models reviewed above, our model also learns by babbling (see Fig. 2A). A learning trial begins by randomly activating the SOM neurons. These activations then propagate via weighted connections to various muscles involved in speech. Using these muscle activation settings, a vocal tract simulator[10] generates a vocalization sound. The acoustic properties of this sound, such as its fundamental frequency (f0) and formant frequencies, can be estimated automatically, or a person can listen to the sound and rate it. We took the automatically estimated fundamental frequency (which can be undefined, often indicating that the sound does not contain voicing) and the first (F1) and second (F2) formant frequencies and set criteria for reinforcement based on these values. We tried several different sets of reinforcement criteria, including the following: (1) no criteria (all sounds were reinforced), (2) sounds must have a defined f0, (3) sounds must have a defined f0 and have F1 and F2 values that are similar to published values for American English speakers, (4) sounds must have a defined f0 and have F1 and F2 vowels that are similar to published values for Korean speakers. Based on the particular reinforcement criteria used in a given simulation, the model's vocalization was either reinforced or not reinforced. If reinforced, then the standard SOM learning algorithm was executed[11], with node activation, which again was randomly determined at the onset of the trial, determining the winning node and the weights from that and neighboring nodes being adjusted to become more similar to the muscle activations just produced. If the vocalization was not reinforced, then no learning took place on that trial.

We evaluated the sounds produced toward the end of the learning period, and compared those sounds across the different reinforcement criteria conditions. These results are shown in Figure 2B and 2C. The reinforcement of all sounds (Condition 1) corresponds to what happens when trying to self-organize around random motor output and as expected results in no increase in defined f0s. In contrast, under all the other conditions, wherein sounds had to have defined f0 in order to be reinforced, we found that toward the end of the learning period, the models produced many more vocalizations with defined f0 compared to early in the learning period. Similarly, when the model was reinforced for producing American English vowels, toward the end of learning the vocalizations it produced were more closely matched to American English vowels compared to when the model was reinforced for producing Korean vowels and vice versa. Taken together, the two results suggest that the reinforcement-modulated learning adaptation

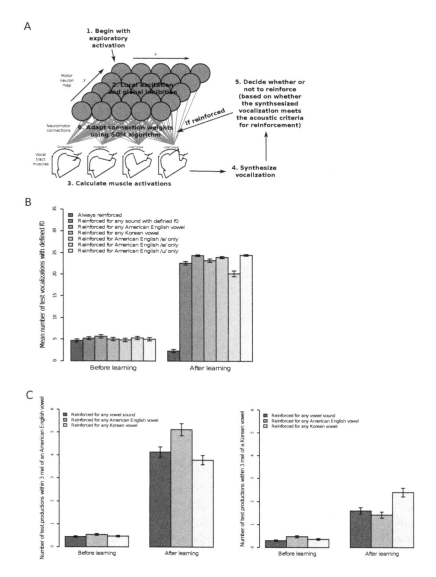

Fig. 2. A: Schematic description of the reinforcement-modulated SOM model[9]. B: Frequency of spontaneous vocalizations having defined f0 early and late in learning when the model is reinforced using various criteria. C: Spontaneous vocalizations' similarity to American English and Korean vowels early and late in learning when the model is reinforced for no particular language, American English-like vowels, or Korean vowels.

to the SOM is approach can work for modeling the development of spontaneous vocalizations coming to increasingly resemble adult speech.

4. Reinforcement-modulated vocal learning with a spiking neural network

In another study, I attempted to use the same principle of reinforcement-modulated learning to model how canonical babbling might come to emerge within infants' spontaneous vocalizations [12]. Around 6–7 months, infants begin to systematically produce canonical babbling, which is the act of producing prespeech vocalizations that have speech-like syllabic timing. Prior to that vocalizations tend to have either sloppier syllabic structure (as in gooing and marginal babbling) or no syllabic structure[1]. It is a bit trickier to model the emergence of canonical babbling than the emergence of phonation or of specific vowel types since babbling is an inherently temporal phenomenon. This may be part of the reason it has not been addressed by any of the other models mentioned above. My approach was to use a spiking neural network, specifically Izhikevich's[13] network, to generate the temporal dynamics of articulatory movement (Fig. 3). The network consists of a thousand neurons, both excitatory and inhibitory, that are connected to each other at random. Each neuron has a voltage parameter that increases under the influence of excitatory neuron input and decreases under the influence of inhibitory neuron input. At 1 ms intervals the voltage of each neuron is calculated and when a neuron's voltage exceeds a particular threshold, the neuron's voltage spikes and then immediately drops back down. When the voltages of all the neurons in the model are summed together, the resulting time series appears rather complex, with oscillations at a range of timescales[14]. By taking a subset of 100 of these neurons and calling them motor neurons, we are thus equipped with a naturally time-varying signal that can be given (after a bit of smoothing) to a vocalization synthesizer as muscle activations. The network learns using a spike-timing dependent plasticity (STDP) mechanism, which can be thought of as a type of Hebbian learning. Izhikevich's model extends previous spiking neural network models by making the rate of STDP, i.e. the rate of learning, dependent on the concentration of the reinforcement-related neurotransmitter dopamine (see also Florian's[15] and Farries and Fairhall's[16] related models). In the babbling model, a human "caregiver" listened to each sound produced by the model and decided whether or not to reinforce it, with the goal of the caregiver being to get the model to produce high-quality canonical babble more often (see Fig. 3B). When the listener decided to reinforce the model, a surge of dopamine was given to the neural network. As can be seen in Figure 3C, the network appears to have learned, through this reinforcement-modulated STDP mechanism, to produce high quality babble, doing so in a rather gradual fashion.

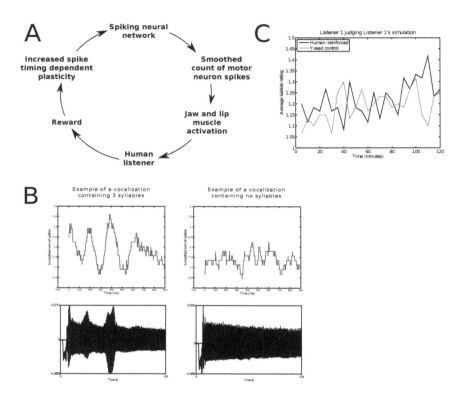

Fig. 3. A: Schematic illustration of the cycle of learning from producing vocalization with a spiking neural network. B: Examples of muscle activations produced by the model (top) and corresponding acoustic waveforms (bottom). The sound on the left would be classified as canonical babble whereas the sound on the right would not. C: Change in the tendency of the model to produce high-quality canonical babble over the course of learning, compared to a control simulation.

5. Conclusions

Taken together, these two models (the SOM model and the spiking neural network model) suggest that reinforcement-modulated learning can be a powerful mechanism for generating learning effects on spontaneous behavior, at least in the vocal domain. It is a general mechanism that can be applied to quite different types of self-organizing neural networks. That said, one might ask whether this reinforcement-gated learning is realistic. From the neurobiological standpoint, there is certainly evidence that dopamine plays a role in increasing STDP during motor learning[17]. And studies of human infant vocal learning indicate, at a behavioral level, that social reinforcement of infant vocalizations increases an infant's rate of vocalization[18] and that in naturalistic contexts mothers do selectively reinforce their infants for vocalizing[19]. Others have proposed that infants might derive intrinsic reinforcement from some of their own actions, such as sounds that expand the infant's current repertoire [20] or stimuli with appropriately moderate complexity[21]; intrinsic reinforcement could perhaps also be generated when infants produce sounds that resemble those they've heard produced by others in their environment. These ideas intrinsic sources of reinforcement still need to be tested in studies with human infants.

Another argument that supports the idea that reinforcement-modulated learning plays a role in this domain is that even deaf infants exhibit many of the changes in spontaneous vocalizations exhibited by hearing infants, such as increase in duration of vocalization, expansion of the ranges of pitches and vocal qualities produced, and even emergence of canonical babbling (although canonical babbling does emerge later for deaf infants)[1]. For profoundly deaf infants, the contagion mechanism of change in spontaneous vocalization modeled by Oudeyer[8] and suggested by Westermann and Miranda[3] is presumably not operational, at least not via the auditory domain, because these infants do not have access to auditory speech input. They do have the ability to feel the vibrations and other tactile sensations generated when they vocalize, and have access to visual and tactile input from other individuals. These tactile and visual sources of information could presumably provide quite useful reinforcement, both social and intrinsic, contingent on the infants' vocal behaviors. Social reinforcement could, as it does for hearing infants, take the form of a touch, smile, look of interest, etc. from a caregiver. Intrinsic reinforcement could result from the deaf infants' deriving more pleasure or interest first of all from vocal motor acts that result in vibration at the larynx and second of all from those that result in sounds with salient rhythmic patterns either at the larynx or through contact between the tongue, lips, and other structures. These social and/or intrinsic reinforcements, coupled with vocal motor exploration and reinforcement-modulated self-organized learning, could explain why deaf infants exhibit many of the spontaneous vocalization advances that hearing infants exhibit.

It is worth saying a few words about why supervised learning approaches are not ideal for this application. Supervised learning methods have been applied in the past in models of speech-language development, and this work has certainly been informative[6,22]. However, in trying to model development of motor learning of the sort that is the focus here, the desired outputs are muscle activations (or other effector values) that we should not assume infants innately know. Using a supervised approach in this context would require knowing ahead of time what the muscle activations or effector values ought to be. In fact, even if the goal is not to model human infants but to control a simulated or robotic vocal tract, it will often be the case that good muscle activations or articulator positions for producing a particular type of sound are unknown. This was the case in both the studies highlighted in this paper—in neither case did the author(s) know ahead of time what muscle activations should be set to in order to create the desired vowel sounds or syllable patterns. Autoencoder networks[23,24] provide a solution to the problem of of unknown targets in perceptual learning problems, so might be expected to also be appropriate for motor learning of the sorts focused on here. Unfortunately this is not the case because of the fundamentally different nature of the learning problem, where, rather than trying to recognize, categorize, and predict sensory inputs, we need to generate outputs that before learning are unknown. Although effects of perception on production are not the focus here, it is possible that autoencoder like processes are involved in infants' perception of their own and others' vocalizations, which could in turn have an impact on vocal motor learning by affecting what infants find reinforcing or by affecting vocal contagion.

It is probably true that multiple mechanisms (particularly contagion and reinforcement-modulated learning) are involved in the changes in spontaneous vocalization observed in human infancy. Thus, an important future direction will be to build models that incorporate multiple mechanisms, and to assess the specific contributions of each and the ways in which different mechanisms might interact. Additionally, although the focus here has been on modeling changes in spontaneous vocalizations, it seems likely that mechanisms such as reinforcement-modulated learning could be integrated nicely into existing models of phonetic learning, such as the DIVA model[7]. Of course, direct comparison to real human infant data will also be important, and it would be helpful to study human infant vocalizations with the goal of seeking more information about these various mechanisms' roles.

Finally, it seems reasonable to expect that reinforcement-modulated self-organized learning might be useful in other action and motor control domains. In fact, the neurobiological work with rats cited above as evidence for reinforcement-modulated motor learning was done in the context of learning a forelimb reaching skill[17]. In modeling work, reinforcement modulation of learning has already been utilized in other action-centered applications such as spatial navigation[25] and decision making[26]. Perhaps it will prove fruitful to pursue a similar approach in modeling language production at higher levels of analysis, such as production of words or sentences, and in modeling additional motor behaviors such as development of limb control for reaching and gesturing.

Acknowledgments

I would like to thank Gert Westermann, Kim Oller, Eugene Buder, Robert Kozma, Mike Goldstein, Thomas Hannagan, Bob French, Pierre-Yves Oudeyer, Clément Moulin-Frier, Mary Beckman, Chris Kello, Jeff Rodny, and the anonymous reviewer for helpful conversations and advice.

References

1. D. K. Oller, *The emergence of the speech capacity* (Lawrence Erlbaum Associates, Mahwah, NJ, 2000).
2. Y. Yoshikawa, M. Asada, K. Hosoda and J. Koga, *Connection Science* **15**, 245 (2003).
3. G. Westermann and E. R. Miranda, *Brain and Language* **89**, 393 (2004).
4. I. Heintz, M. Beckman, E. Fosler-Lussier and L. Ménard, Evaluating parameters for mapping adult vowels to imitative babbling, in *Proceedings of the 10th Annual Conference of the International Speech Communication Association (INTERSPEECH)*, (Brighton, UK, 2009).
5. A. S. Warlaumont, G. Westermann and D. K. Oller, Self-production facilitates and adult input interferes in a neural network model of infant vowel imitation, in *AISB 2011 Computational Models of Cognitive Development*, eds. D. Kazakov and G. Tsoulas (Society for the Study of Artificial Intelligence and the Simulation of Behaviour, York, UK, 2011).
6. H. Kanda, T. Ogata, T. Takahashi, K. Komatani and H. Okuno, Continuous vocal imitation with self-organized vowel spaces in recurrent neural network, in *2009 IEEE International Conference on Robotics and Automation*, 2009.
7. F. H. Guenther, S. S. Ghosh and J. A. Tourville, *Brain and Language* **96**, 280 (2006).
8. P.-Y. Oudeyer, *Journal of Theoretical Biology* **233**, 435 (2005).
9. A. S. Warlaumont, G. Westermann, E. H. Buder and D. K. Oller, *Neural Networks* **38**, 64 (2013).

10. P. Boersma, *Functional phonology: Formalizing the interactions between articulatory and perceptual drives* (Holland Academic Graphics, The Hague, 1998).
11. T. Kohonen, *Proceedings of the IEEE* **78**, 1464 (1990).
12. A. S. Warlaumont, A spiking neural network model of canonical babbling development, in *Proceedings of the 2012 IEEE International Conference on Development and Learning (ICDL)*, 2012.
13. E. M. Izhikevich, *Cerebral Cortex* **17**, 2443 (2007).
14. C. T. Kello, J. Rodny, A. S. Warlaumont and D. C. Noelle, *Critical Reviews in Biomedical Engineering* **40**, 501 (2012).
15. R. V. Florian, *Neural Computation* **19**, 1468 (2007).
16. M. A. Farries and A. L. Fairhall, *Journal of neurophysiology* **98**, 3648 (2007).
17. K. Molina-Luna, A. Pekanovic, S. Röhrich, B. Hertler, M. Schubring-Giese, M.-S. Rioult-Pedotti and A. R. Luft, *PLoS ONE* **4**, p. e7082 (2009).
18. M. H. Goldstein, A. P. King and M. J. West, *Proceedings of the National Academy of Sciences of the United States of America* **100**, 8030 (2003).
19. J. Gros-Louis, M. J. West, M. H. Goldstein and A. P. King, *International Journal of Behavioral Development* **30**, 509 (2006).
20. C. Moulin-Frier and P.-Y. Oudeyer, Curiosity-driven phonetic learning, in *Proceedings of the 2012 IEEE International Conference on Development and Learning (ICDL)*, 2012.
21. C. Kidd, S. T. Piantadosi and R. N. Aslin, *PLoS ONE* **7**, p. e36399 (2012).
22. M. S. Seidenberg and J. L. McClelland, *Psychological Review* **96**, 523 (1989).
23. D. Mareschal, R. M. French and P. C. Quinn, *Developmental Psychology* **36**, 635 (2000).
24. G. E. Hinton, *Trends in Cognitive Sciences* **11**, 428 (2007).
25. H. Voicu, R. Kozma, D. Wong and W. J. Freeman, *Connection Science* **16**, 1 (2004).
26. R. Ratcliff and M. J. Frank, *Neural Computation* **24**, 1186 (2012).

A COMPUTATIONAL MODEL OF THE HEADTURN PREFERENCE PROCEDURE: DESIGN, CHALLENGES, AND INSIGHTS

C. BERGMANN

Centre for Language Studies and International Max Planck Research School for Language Sciences, Radboud University, P.O. Box 9103, 6500 HD Nijmegen, The Netherlands
E-mail: CBergmann@science.ru.nl

L. TEN BOSCH and L. BOVES

Centre for Language and Speech Technology, Radboud University, P.O. Box 9103, 6500 HD Nijmegen, The Netherlands

The Headturn Preference Procedure (HPP) is a frequently used method (e.g., Jusczyk & Aslin [1]; and subsequent studies) to investigate linguistic abilities in infants. In this paradigm infants are usually first familiarised with words and then tested for a listening preference for passages containing those words in comparison to unrelated passages. Listening preference is defined as the time an infant spends attending to those passages with his or her head turned towards a flashing light and the speech stimuli. The knowledge and abilities inferred from the results of HPP studies have been used to reason about and formally model early linguistic skills and language acquisition. However, the actual cause of infants' behaviour in HPP experiments has been subject to numerous assumptions as there are no means to directly tap into cognitive processes. To make these assumptions explicit, and more crucially, to understand how infants' behaviour emerges if only general learning mechanisms are assumed, we introduce a computational model of the HPP. Simulations with the computational HPP model show that the difference in infant behaviour between familiarised and unfamiliar words in passages can be explained by a general learning mechanism and that many assumptions underlying the HPP are not necessarily warranted. We discuss the implications for conventional interpretations of the outcomes of HPP experiments.

Keywords: Language Acquisition, Segmentation, Headturn Preference Procedure, Attention

1. Introduction

The number of studies on infants' linguistic abilities has drastically increased over the past decades. The work by Jusczyk and Aslin [1] serves as a seminal study. In a series of experiments using the Headturn Preference Procedure (HPP) the authors investigated the abilities of 7.5-month-olds to memorise and recognise words when they are spoken in isolation or embedded in continuous sentences. Jusczyk and Aslin conclude from their results that infants can extract, memorise, compare and recognise words; and that the words are represented with sufficient detail to prevent confusion with similar words or mispronunciations.

HPP studies are usually split into two parts: First the infant is familiarised with e.g., words for a set amount of time. Subsequently, the response to passages that contain either familiar or novel words is observed. The measurement of infant responses during HPP studies is the time an infant spends listening to the passages, indicated by the time the head is turned towards a flashing light and a loudspeaker, installed at the left and right side of the infant. When the infant shows disinterest by turning away from the flashing light (and the speech stimuli) for more than two consecutive seconds the presentation of a trial ends prematurely. A listening preference for passages containing familiar words is considered to indicate recognition of the known words, even when they are embedded in sentences.

Behavioural paradigms such as the HPP aim to tap into infants' linguistic abilities by inferring cognitive processes from observable behaviour, which requires assumptions that cannot be tested directly. Linking overt behaviour in HPP experiments to underlying cognitive processes is based on three main assumptions. First, the length of time during which the infant's head is turned toward a flashing light and a loudspeaker is considered a measure of interest that results from the processing of acoustic stimuli. Second, longer listening times for passages containing previously familiarised words presumably show that these words are segmented from the sentences and subsequently 'recognised'. Third, individual differences in infants will not affect the overall outcome of an experiment, as the main comparison (listening to novel or familiar test stimuli) takes place within participants. These assumptions cannot be tested easily in experiments with infants. Thus, we developed a computational model of the HPP which not only requires explicit specification of all assumptions but also allows us to investigate whether our model can accurately simulate the behaviour typically observed in infant studies.

Our model subscribes to the first assumption, but it makes all processing explicit, and it allows for an investigation of the second and third assumption. Specifically, our model does not decode and memorise speech in the form of sequences of discrete phonemes, an ability that infants are still in the process of acquiring [2,3], nor does it segment speech into discrete, word-like units. Instead, we assume that episodic, possibly holistic, representations are at play [4] that can be formed by means of general perceptual and cognitive capabilities. The input to the model consists of real speech, which is encoded via a language-independent auditory processing step. The encoded input is then analysed in terms of representations of the familiarised words that are stored in the model's memory. Passages that do not contain familiarised words are processed in the exact same way, but the results may be different from processing passages with familiarised words. The model contains a module that converts the processing results into simulated headturns, again using only general-purpose capabilities. During this conversion we can investigate the third assumption with a focus on attention span.

2. The Headturn Model

The architecture of the model is shown in Fig. 1. Computing whether a test stimulus is similar to previous experience requires mechanisms to process auditory input, to store 'old' information that has been experienced within the past few minutes (at least for the duration of the experiment), to compare new stimuli to stored representations in the same format, and to generate behavioural responses.

Acoustic preprocessing is employed to transform variable-length speech input (words spoken in isolation and complete sentences) into a fixed-length vector. For this purpose, we use a Histogram of Acoustic Coocurrences (HAC, [5]). The histogram counts acoustic events (based on Mel-Frequency Cepstral Coefficients, henceforth MFCCs) with a sampling rate of 10 ms and coocurrences of these events at distances of 20 and 50 ms. To represent speech in enough detail for the subsequent recognition task, hundreds of possible acoustic events are distinguished. Building the HAC representation does not invoke language specific knowledge (e.g., about phonemes).

Using the HAC encoding, we create an **internal memory** that represents what an infant brings to the test phase of a typical HPP experiment. The internal memory in our model consists of two parts. First, to model the infant's exposure to the ambient language before the lab visit, we randomly selected utterances from a corpus of infant-directed speech [6]. The second part contains the familiarised stimuli that the model was exposed to during the first phase of an HPP experiment.

To compare each test stimulus to the content of the internal memory, a **matching procedure** is necessary. Non-negative Matrix Factori-

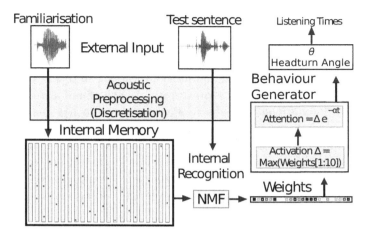

Fig. 1. The Headturn Preference Procedure model, with processing stages and flow of information from external input (top left) to overt behaviour in an experimental setting (top right).

sation (NMF, [7]) assumes that a new token can be approximated as a positive weighted sum of all stored representations. Interestingly, NMF can be phrased in the same terms as activation and inhibition in neural networks [8], but it can be applied to the hand-crafted memory used in the present model.

The variant of NMF in the present paper minimises the Kullback-Leibler divergence between a HAC-encoded novel test stimulus and its approximation as a positive weighted sum of the contents of the complete internal memory. NMF decoding of an unknown utterance results in a vector with positive weights for all representations stored in the memory. The weights are available immediately after the end of an utterance. To allow for comparisons between the decoding of different utterances, the weights are normalised to sum to one. A **familiarity score** represents how much the familiarised tokens contribute to the approximation of a test stimulus. In the experiments described in this paper we enforce a familiarity preference by basing the score on the highest weight in the section of the internal memory that contains the familiarised token.

In HPP studies, the headturns of an infant are measured as an overt sign of underlying attention to the speech stimuli. To simulate the attention of an infant to the test stimuli, we transform the discrete-time familiarity scores of the test utterances into a continuous function. Since we assume that the familiarity score is available immediately after the end of the utterance and we know the duration of all utterances, the discrete familiarity scores can be converted to Dirac pulses $\delta(t_i)$ with an amplitude a_i equal to the familiarity score of the ith test utterance, separated by the duration of the utterances. The sequence of pulses $a_i \cdot \delta(t_i)$ is converted into a continuous function by applying an exponential decay, resulting in the **attention function** $a_i \cdot e^{-\alpha t}$, where α is a (positive) parameter specifying the decay rate, and t denotes time (c.f. Fig. 2 for illustration). The exponentially decaying attention function can be interpreted as the degree to which the head is turned towards the flashing light and the loudspeaker. While the function value is high, the infant's head is completely turned towards the flashing light. As the attention value decreases, the head is more likely to be turned away from the lamp.

During the HPP, the experimenter interprets the angle of the head relative to the center and side lamps in terms of discrete states. In the model, the experimenter's decisions are implemented as a Finite State Machine (FSM). The FSM takes the continuous attention function as input and calculates the listening time (for each paragraph of six sentences). To that end, the FSM uses a **threshold** θ, which corresponds to the angle of headturn necessary to be considered a sign of attention. As long as the attention value exceeds θ, the head is turned sufficiently in the direction of the flashing light. As soon as the attention level drops below θ, the experimenter considers the infant to not be listening to the speech stimuli, as indicated by an angle of the headturn that is too far away from the lamp. If the value of the attention function stays below θ for more than two consecutive seconds, the trial is terminated (as in HPP studies).

The parameter $\rho \geq 0 + \theta$ models the attention level at the start of each test trial consisting of a six-sentence paragraph. It can be conceptualised as the initial degree of interest in the flashing lamp, before onset of the additional speech stimuli. The value is defined relative to the threshold θ, since a trial only starts when the infant's head is turned towards the flashing lamp sufficiently to be considered interested.

3. Experiments

We first investigated how well the model distinguishes sentences containing novel words from those with familiar words. The possibility that the internal ability to differentiate the two is speaker-dependent was examined by repeating all simulations using recordings from four native speakers of British English (two female), with one speaker for each set of experiments.

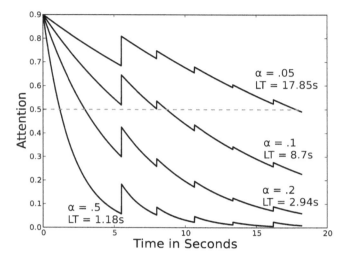

Fig. 2. Exemplar attention functions based on familiarity values and sentence durations of Speaker 01. The threshold θ is set to .5, listening times (LT) across exemplary values for α are depicted. In all cases the initial attention level is .9, which exceeds the threshold θ (a necessary requirement for starting a trial in an HPP experiment).

Subsequently, we addressed the behaviour-generating parameters α, the attention decay, and θ, the assessment criterion. If the model is internally able to distinguish between known and unknown words, even if they are embedded in sentences, how does this translate into observable behaviour as measured by listening time differences?

The ranges for all behaviour-generating parameters were chosen to lie between 0, an unrealistically low value, and a cognitively motivated upper limit. The maximum value for α was set to .5, which corresponds to a steep attention decay function. Higher values of α will most likely lead to no observable differences in listening times as attention decays too fast across conditions to be recovered. The parameter ρ was fixed at $.4 + \theta$, which simulates a sufficient initial inclination to turn the head to the side lamp. For θ the maximum value was set to 2. The larger θ is, the shorter the time during which the experimenter considers the infant head being turned sufficiently towards the side. A too strict threshold will rarely consider the headturn to be sufficient and a threshold near 0 simulates an experimenter who always considers the head turned towards the lamp.

Differences in listening times to passages of six sentences containing either novel or familiar words were the dependent measure to model HPP experiments. We generated 30 test passages for each of the four words to obtain a reliable measure of model performance. To this end, we repeatedly chose random sets of six sentences out of the 24 available sentences per test word. The listening time differences obtained with the 30 test passages were averaged and form the basis for the analysis of the impact of the parameters α and θ in the process that converts the familiarity scores into listening times.

3.1. *Material*

A range of acoustic stimuli are necessary to simulate headturn studies: words spoken in isolation, embedded in continuous sentences, and additionally some unrelated speech stimuli to model past language experience. We selected words and sentences spoken by native speakers of British English and recorded in a virtually noise-free environment [6]. Four adult speakers were available for the present study, two of which were female. Some basic acoustic descriptions can be found in Table 1.

For each familiarised word stored in memory, five different tokens of two monosyllabic words were available (*frog* and *doll* or *duck* and *ball*; selected to be similar to the original stimuli of Jusczyk and Aslin [1]). As test stimuli we randomly selected 24 short sentences for each of the four words. These sentences were identical for all speakers and the same test sentences could be novel and familiar stimuli, depending on the familiarisation condition. Familiarisation stimuli and test sentences all came from the same speaker within every simulation. When the speaker changes between familiarisation and test, infants seem to sometimes lose the ability to recognise familiarised words embedded in sentences [9,?]. In the simulations the internal memory consisted of 111 HAC vectors, 10 containing 5 tokens for each of the two familiarised words, 100 containing sentences spoken by the same speaker that did not contain one of the four target words, and one containing background noise (silence obtained during the recording session). The 101 non-target slots were identical in all simulations within speaker.

4. Results

4.1. *Familiarity Scores*

As a first step, we investigated whether the internal familiarity scores can differentiate between test sentences containing either novel or famil-

Table 1. Acoustic properties of all speakers: pitch (mean value in Hertz) and average sentence duration in seconds

	Mean Pitch (Hz)	Sentence Duration (s)
Speaker 01 (m)	121.66	2.70
Speaker 02 (f)	180.80	2.57
Speaker 03 (m)	166.63	2.64
Speaker 04 (f)	180.58	2.88

iar words. The familiarity scores can be computed independently of the three parameters α, θ and ρ, all of which affect the conversion of the familiarity scores to listening times. Per speaker, we calculated the mean and standard deviation of the familiarity score for all sentences in the test set. To test whether the familiarity scores (which we do not assume to be normally distributed) differ significantly between the two test conditions, we subjected the scores for all sentences (96 sentences, each being used once as a novel and once as a familiar test item) to a Mann-Whitney U-test. Both the descriptive and analytic statistics are shown in Table 2.

Table 2. Mean and standard deviation for familiarity scores; Mann-Whitney U score and p-value per speaker

Speaker	Familiarity Score		U-score	p-value
	familiar	novel		
Speaker 01	.049 (.029)	.036 (.025)	3845	.0238
Speaker 02	.057 (.028)	.050 (.033)	3537	.0027
Speaker 03	.108 (.052)	.089 (.044)	3922	.0374
Speaker 04	.087 (.042)	.071 (.039)	4327	.2331

Table 2 shows that the model's internal familiarity scores are consistently higher for sentences that contain known words. However, there are considerable discrepancies between the four speakers. Most of the coefficients of variation (i.e., standard deviations divided by the mean) are < 0.5, which implies that there are also substantial divergences between the individual sentences in a set. For Speaker 04, a female speaker, the scores of familiar and unknown words is not significantly different, mainly due to the relatively large standard deviations in comparison to the mean, leading to a high coefficient of variation.

4.2. *Behaviour-Generating Parameters*

Although the familiarity scores for known and unknown words did not differ significantly for Speaker 04, we included all speakers in this analysis for two reasons: First, the familiarity scores of Speaker 04 were still numerically higher for known words. Second, it is interesting to investigate whether it is possible to find reasonable values of the parameters α and θ to see a familiarity preference emerge even when the stimuli are spoken by Speaker 04. Therefore, we inspected the listening time preferences for all speakers for plausible ranges of the parameters α and θ. In previous research [10] it has been shown that the behaviour-generating parameters strongly affect the strength and even the presence of an observable listening time difference.

For all speakers (including Speaker 04) the model is able to generate a robust listening time difference (amounting to several seconds per paragraph) for some parameter combinations. The maximum value of this familiarity advantage in listening time ranged from 1.95 seconds for Speaker 01 to 2.6 seconds for Speaker 03. In comparison, the average listening time advantage for passages containing familiar words obtained by Jusczyk and Aslin [1] amounts to 1.25 seconds in Experiment 1, where the passages had a similar duration to those our experiments.

The impact of the behaviour-generating parameters on the preference in listening to familiar versus novel test stimuli is illustrated in Fig. 3, which shows the listening time differences as a function of the two parameters α and θ for Speaker 01. A robust listening time difference can be obtained over a number of parameter settings. However, if the threshold θ or the attention decay parameter α are set to very low values, the model spends the whole paragraph with attention above threshold, irrespective of whether the test items contain known or novel words. If the threshold θ is set too high, the attention level will soon drop below the threshold for more than two seconds, again irrespective of the contents of the test utterances. The same is true if the attention decay parameter α is set too high.

The impact of quickly waning attention, modelled by high values for α, or of a strict assessment criterion θ, can be counteracted by adjusting the respective other parameter. Thus, even for a modelled very fast attention decay with $\alpha = .5$, a preference in listening time with a sufficiently adjusted threshold θ can be obtained.

134

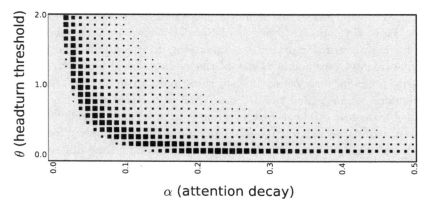

Fig. 3. Listening time differences for Speaker 01 to familiar versus novel stimuli across different values for α and θ. The maximal square size corresponds to a listening time advantage of 1.95s per paragraph for familiar stimuli. The initial attention is $.4 + \theta$.

5. Discussion

In the present paper we have described and assessed a computational model that simulates a frequently-used behavioural paradigm in infant language acquisition research. The Headturn Preference Procedure (HPP) indirectly measures cognitive skills via listening times, indicated by headturns, to auditory stimuli. The present model simulates infant listening behaviour to allow for a direct comparison with HPP studies. However, we are also able to inspect the internal processes and investigate the model's underlying abilities that give rise to the simulated listening behaviour. The model implements only general-purpose abilities and allows us to examine two assumptions that are usually present in HPP studies. A preference for listening to passages that contain familiarised words presumably reflects an ability to segment those words from the sentences, and to subsequently recognise these words. We find that no explicit, dedicated segmentation procedure is necessary to simulate recognition of known or unknown words, even when they are embedded in continuous sentences. Individual differences across infants are assumed to not affect the overall measured recognition, as indicated by listening preference to familiar over novel test stimuli. Our results show that attention span can lead to listening times that do not reflect internal abilities to recognise familiar words embedded in sentences.

In assessing the model in detail we first tested whether speaker-specific properties can affect the model's internal ability to reliably distinguish sentences containing a familiar word from sentences for which this is not the case. The model's ability to generate different familiarity scores for both conditions depended, to some extent, on the speaker who recorded the familiarisation and test stimuli. The exact sources of the between-speaker variability remains unclear, but it must lie in the acoustic properties of the single speakers, as the sentences, the HAC-encoding and the matching procedure all were held constant. Therefore, the choice of the speaker in HPP experiments seems to have a non-negligible impact on experimental outcomes. However, it cannot be excluded that this phenomenon is a model-specific property, since the HAC encoding based on MFCCs is only one possible way to generate internal episodic representations.

Investigating listening preferences, the dependent measure in HPP studies, we found that the model is able to generate data comparable to infant performance, even when the internal familiarity scores could not reliably distinguish between novel and familiar test items. However, the two parameters that govern the generation of listening times, the attention decay α and the assessment threshold θ strongly affect whether the observable responses can distinguish between passages containing novel or familiar words - irrespective of the underlying ability to reliably do so. This finding suggests that in HPP studies factors not directly related to internal cognitive abilities can affect the outcome. More precisely, both infant-specific properties (attention span) and experimenter-dependent criteria (the assessment threshold), can interact to either bring out or obscure underlying differences in the responses to test stimuli. For all speakers, independent of the degree of separation of their internal familiarity scores in response to novel or familiar test items, we replicated longer listening times to familiar test stimuli for some settings of the parameters α and *theta*. The finding that a statistically reliable ability to distinguish test conditions internally does not affect overt behaviour substantially suggests that, under ideal conditions, the HPP seems to be sensitive to subtle processing differences.

In summary, we present a computational end-to-end model of the HPP, a frequently used behavioural measure of infants' abilities to process language. We replicated infant behaviour and explored which underlying conditions are sufficient to do so. We investigated three main assumptions that are in place during HPP studies. The link between overt behaviour, head-turns, and internal processing was taken over from infant studies. Addressing the assumption that infants segment words from continuous speech prior to recognition, we found that a general matching procedure based on words and complete test utterances, without prior segmentation, proved sufficient for this task. Finally, we show that individual infant characteristics can obscure cognitive abilities. Future research is necessary to investigate the adverse effect of a too short attention span.

Acknowledgements

The research of Christina Bergmann is supported by grant no. 360-70-350 from the Dutch Science Organisation NWO.

References

1. P. W. Jusczyk and R. N. Aslin, *Cognitive Psychology* **29**, 1 (1995).
2. P. K. Kuhl, *Nature reviews neuroscience* **5**, 831 (2004).
3. R. S. Newman, *Current Directions in Psychological Science* **17**, 229 (2008).
4. S. Goldinger, *Psychological review* **105**, 251 (1998).
5. H. Van hamme, Hac-models: A novel approach to continuous speech recognition, in *Ninth Annual Conference of the International Speech Communication Association*, 2008.
6. T. Altosaar, L. ten Bosch, G. Aimetti, C. Koniaris, K. Demuynck and H. van den Heuvel, A speech corpus for modeling language acquisition: Caregiver, in *Proceedings of the International Conference on Language Resources and Evaluation (LREC), Malta*, 2010.
7. D. Lee and H. Seung, *Nature* **401**, 788 (1999).
8. H. Van hamme, On the relation between perceptrons and non-negative matrix factorization, in *Signal Processing with Adaptive Sparse Structured Representations Workshop*, 2011.
9. D. M. Houston and P. W. Jusczyk, *Journal of Experimental Psychology: Human Perception and Performance* **26**, 1570 (2000).
10. C. Bergmann, L. Boves and L. ten Bosch, A model of the headturn preference procedure: Linking cognitive processes to overt behaviour, in *IEEE International Conference on Development and Learning and Epigenetic Robotics (ICDL)*, 2012.

RIGHT OTITIS MEDIA IN EARLY CHILDHOOD AND LANGUAGE DEVELOPMENT: AN ERP STUDY

M. F. ALONSO[*]

AIDA. Communication and language department
Santiago de Chile

P. UCLÉS[†]

Hospital Miguel Servet, Neurophysiology service.
Zaragoza, Spain

P. SAZ[‡]

Universidad de Zaragoza, Medicine.
Zaragoza, Spain

The aim of this study is to analyze the effects of right otitis media on language development in the first three years of a child's life. A study of middle-latency evoked potentials was carried out to observe the correlation between right otitis media history and language disorders. We tested two groups of children, both aged between 3 and 6.11 years old. The first group had a history of right otitis media in the earliest years of life (n=12) and the second group had a history of left otitis media, had not suffered otitis or had suffered from otitis but made a full recovery (n=17). Several battery language tests were applied to assess the children's language abilities. Auditory brainstem-evoked responses and middle-latency evoked potentials were employed to establish if plastic changes had occurred after prolonged periods of hearing loss in the auditory path and the cerebral cortex. The fact of being a case is related to lower scores in phonetic, phonological and morphosyntax tasks. There is a correlation between the phonetic and phonological system and the auditory brainstem evoked response with the energy of the middle latency response.

---

[*] Corresponding author: ma.francisca.alonso@gmail.com; Tel no: 56 – 09 – 66793870
[†] pucles@salud.aragon.es; Tel no: 34 – 976765500
[‡] pablosaz@unizar.es; Tel no: 34 – 976761000

1. Introduction

1.1. *Language and auditory pathway development*

The main mode of communication is the spoken language and this is based primarily on phonemic dynamic auditory processing. The basis of the language development, the process of maturation and self-programmed learning, occurs in the critical period of growth.[1] This period is characterized by changes such as neuronal increase and growth, chemical changes in the cerebral cortex and electrophysiological modifications. The myelination of the central acoustic pre-thalamic pathway is completed in the first year; the post-thalamic acoustic pathway has a slower myelogenous cycle, which takes the first two years. It is also in these years the peak of cerebral plasticity.[2] This period is critical because the dynamic auditory process is developing and the basis of formant recognition is generating. We propose that any alteration of the signal as this stage distorts the frequency perception, modifying the boosting and specialization of cortex areas.

The areas involve in language are quite close to the primary auditory cortex responsible of the sound perception and recognition. But the auditory functions of the temporal lobes are not identical; the left temporal lobe specializes in sequential processing and linguistics inputs, and the right temporal lobe in the melodies, ambient noise and language prosody in right-handed individuals.[3] Recently a frequency recognition area had been identified right next to the auditory cortex, this area involve the formant analysis to distinguish the acoustical patterns on each phoneme.[4]

The right ear pathway is the primary auditory route that reaches the language centre, and specifically with the frequency recognition area. The specialization of the temporal lobes and the functional dominance of the contra-lateral auditory pathway draw attention to the possible implications of the interaction between these systems. A mice study showed that sensory deprivation, with right ear blocked, in critical periods of brain development and high cerebral plasticity caused changes in the cerebral cortex function.[5]

1.2. *Middle ear infection: Otitis media*

Otitis media (*Latin*) is an ear infection and implies a middle ear obstruction. Otitis media (OM) is the second most common cause of children disease. It is the presence of inflammation, infection and/or exudates in the middle ear cavity and it is frequently generated by common bacteria related to the upper airway.[6,7] This pathology can occur since the first month of life. A 13% of three months old infants, 60% of one year old children and 80% of four years old children had suffered an OM.[8] The high incidence in children is due to the shorter and horizontally angulated Eustachian tube. Consequently they have a poor drainage of bacteria and pus. Since the middle ear is swollen (Acute OM), full of fluids (OM with effusion) or without tympanic membrane (Chronic OM) the wave sound is attenuated (Figure 1). The acute OM show important pain and minimum hearing lost, the effusion OM is generally with hearing lost but without pain and after some relapse appears the Chronic OM also with important hearing lost. The vibration that reaches to the hair cells, through the ossicles and the oval window it is already modified in its frequency spectra, mainly in the low frequencies due to the mass effect. This pathology involves a temporary reduction in the conduction of sound through the middle ear, generating a temporary auditory loss of between 15 decibels (dB) and 40dB.[9]

1.1   1.2   1.3

Fig. 1. Tympanic membrane view in three types of OM. 1.1: Acute OM with swollen eardrum; 1.2: OM with effusion full of fluids; Chronic OM with a wide perforation.

1.3. *Otitis media and language development*

In a longitudinal study of children who had effusive OM in the first three years of life, it was observed that these had significantly worse hearing than those individuals who had not suffered otitis. The study emphasized that this fluctuation of hearing loss caused by the infection leads to inconsistencies in the stimulation of the central auditory system.[10] Fluctuations in hearing thresholds in the first three years of life, difficult the language acquisition process, considering that it is in this period that children refine the ability to discriminate sounds, especially in their mother tongue.[11]   Several studies[12,13,14,15] have reported a deficit in stimuli categorization, in subjects that have suffered OM in the critical period of the development. Nittrover[16] observed that children with early OM history experienced changes in phonological awareness.[17] Menyuk detected a correlation between long periods of OM in the first three years and phonological and morphosyntax alterations.[11] Also phonological deficits may affect semantic and morphosyntax levels[18], based on the fact that if an individual does not discriminate the formants of the signal he or she cannot develop a strong semantic database. These alterations affect reading fluency and comprehension. Some authors affirm that the influence of otitis media on binaural hearing declines with age.[19,20] Interestingly, there are some studies suggesting that otitis media is not related to language development.[21,22,23] These studies do not disclose which type of otitis was considered, if were unilateral or even if the otitis media was considered as an independent variable. A study of long-term otitis media in children[24] established that auditory processing disorders have a central origin due to the occurrence of alterations in otitis-free periods. It is believed that language impairment is a consequence of phonological deficits in short-term memory and a deficit in the temporal ability of the decoding resolution. Bishop argues that this change affects the ability to discriminate certain frequencies, distorting phonological categorization.[25]

The aim of this study is to analyze whether the right otitis media, during the first three years of age, disturb the main auditory pathway that reach to the perisylvian language centre and witch are the implications in the language development, specifically in the frequency recognition.

2. Material & subjects

A study was perform with patients diagnosed with chronic right dominant or right unilateral OM during the first three years of life, as documented in their clinical history. The recruitment of patients was carried out in the children's otolaryngology (ORL) service of Hospital Universitario Miguel Servet. The inclusion criteria were as follows: the patients had to be aged between 3 years and 6 years and 11 months and have a clinical history of right OM in the first three years of life. The exclusion criteria were that they suffered other pathologies or syndromes. It was not possible to verify the duration of each OM episode of every participant, but only patients suffering from chronic otitis were included in the study. The degree of hearing loss of each episode data was estimated using a tuning fork (Rinne/Weber) in most cases and with immittance measures and tonal audiometric in others. Additionally, a control group composed of children of the same age was examined. The control group consisted of children who had not experienced otitis media during the first three years of life, or who had suffered a single episode from which they had fully recovered. Also a group of left chronic otitis media was examined. All the participants had an informed consent form signed by their parents. The ethics committee of the Instituto Aragonés de Ciencias de la Salud approved this project.

2.1. *Language assessment*

A linguistic evaluation was carried out by applying standardized tests and/or developmental stage assessment for each level. For the phonetic–phonological level, the following variables were analyzed: perceptual and praxis skills, phonological simplification process, voluntary phoneme production, phonological resolution, phonological awareness and faulty articulation points.[26] The Monfort intelligibility test was used for the evaluation of children´s phonological development and TAR (Repeated Articulation Test) for the phonetic level. In the morphosyntax level the comprehension and expression of syntactical and grammatical structures were analyzed as variables. Aguado´s Test (TSA) was used for the assessment of the child's morphosyntax development.[27] The semantic assessment focused on word recognition using Dunn's PPVT (Peabody Picture Vocabulary Test) images.[28] Screening was performed to rule out pragmatic developmental disorders but it was not considered as a variable for this study. These tests were selected because they provide a comprehensive assessment of language levels skills and have been standardized for the Spanish population.

2.2. *Auditory perception assessment*

A hearing assessment was carried out using auditory brainstem-evoked responses (ABR) and middle latency evoked responses (MLR).[29,30] Electrodes recorded the neuro-electrical activity of the auditory nerve, central auditory pathways, subcortical and cortical areas (Figure 2). The right ear was assessed first with masking noise in the left ear, then vice-versa. The sample frequency was 547 Hz and in 100 ms there was an average of 1024 sweeps and the waveforms were stored on the hard disk of Medtronic KeyPoint (Denmark A/S) equipment for further analysis. Considering that the gamma frequency, between the 30 and 60 Hz, is involved in sound perception[31], the variables were the amount of energy measured in two time windows: between 3 and 6.5 milliseconds related to the brainstem and between 15 and 30 milliseconds correlated with the cerebral cortex.

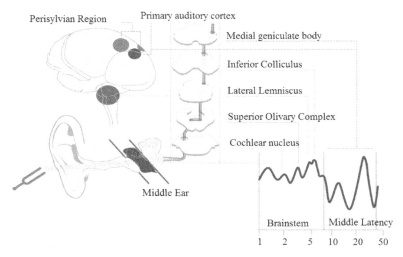

Fig. 2. Auditory pathway and it correlation with the auditory brainstem-evoked responses.

This procedure allow us to assess the sound frequency perception with differentiation of the cochlea and cortex tonotopic distribution and without the involve attention, memory or higher cognitive process.

2.3. *Analysis method*

Language scores were analyzed according to Spain´s age-standardization and percentiles were obtained for their position in relation to the general population.

The Daubechies´s wavelet was used for the analysis of the ABR and MLR in a time-frequency scale method (Figure 3) in MatLab. The decomposition of the bio-electric signal evoked potential was performed and the gamma-frequency band which matched scale point nine was located. It was obtained in time-scale two time windows: 3 to 6.5 milliseconds (ABR) and 15 to 30 milliseconds (MLR), which correspond to the auditory brainstem way and the cerebral cortex respectively.

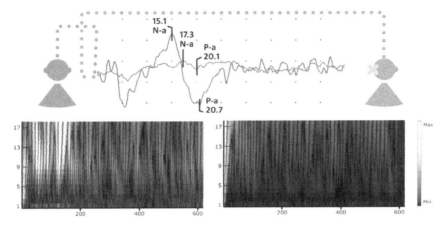

Fig. 3. We analyze the differences in the neural oscillation patterns between 30 and 60 Hz, the gamma band. The Daubechies's wavelet allow a time/frequency/amplitude exploration.

The results were analyzed using "Statistical Package for the Social Science" (SSPS) software. The qualitative variables are presented by frequency distribution of percentages in each category.[32] Quantitative variables indicated central tendency (mean or median) and dispersion (standard deviation or percentiles). The Mann Whitney-U and Spearman test were applied.[33] Normality was assessed with Kolmogorov-Smirnov and was compared using Shapiro-Wilk test.[34] Finally a correlation was made between the energy of the evoked responses found at the two windows and the different language level scores using the Spearman rho coefficient.

3. Results

Assessments were carried out on 29 eligible candidates, from October 2010 to February 2011, 41% (n = 12) belonged to the case group and 58% (n = 17) to the control group. The average age of cases was 4 years 9 months and controls 4 years 9 months (Table I). All participants had Castilian as their mother-tongue and had an average educational background. The hearing status at the time of assessment was without acute otitis, but with chronic consequences of previous otitis.

Table 1. Baseline characteristics

Case/Control	n participants	Age
Right OM	12	4 years 9 month
Control	17	4 years 9 month
Control/ No-OM	14	4 years 6 month
Control/ Left OM	3	5 years

After obtaining the results from each of the tests a statistical analysis was carried out using SPSS. Due to the small sample size and non-normal distribution of quantitative variables, non-parametric statistical tests were used: U de Mann Whitney and Spearman's correlation coefficient.[32]

The left OM group showed homogeneous results with the no-OM control group, and because of the size sample, were merged into a single control group. The descriptive analysis shows the mean, standard deviation, median and range of groups of cases and controls for each level of language. This is shown in table II, which describes the frequency measurements.

Table 2. Frequency measures

	Case				Control			
	Mean	SD	Median	Range	Mean	SD	Median	Range
Age	4.99	0.89	5.08	2.83	5.13	1.22	5.25	3.92
Phonetic	1.17	0.83	1.00	2.00	0.24	0.44	0.00	1.00
Phonology	3.75	2.01	4.00	6.00	0.53	1.01	0.00	3.00
Morphosyntax	44.17	18.69	37.50	55.00	74.41	9.66	75.00	35.00
Semantic	52.50	7.54	50.00	20.00	58.24	7.28	60.00	20.00
ABR μV^2	6.35	9.31	3.91	34.26	7.54	6.33	6.45	19.44
MLR μV^2	20.84	16.07	16.61	45.30	44.74	39.92	36.76	117.92

Subsequently, an analysis of normality was carried out through Kolmogorov-Smirnov[35]; a non-parametric test was used to determine the goodness of fit of two probability distributions in relation to each other. The Shapiro Wilk test[34] was used to contrast the normality of a data set. Table III shows the results obtained using these tests for each of the tests.

Table 3. Normalcy Analysis where: [a] Normal distribution and [b] Non- normal distribution

		Kolmogorov-Smirnov		Shapiro-Wilk	
		Statistic	Sig.	Statistic	Sig.
Age[a]		0.115	0.200	0.959	0.315
	Case	0.133	0.200	0.963	0.825
	Control	0.150	0.200	0.939	0.312
Phonetic[b]		0.340	0.000	0.733	0.000
	Case	0.258	0.027	0.802	0.010
	Control	0.469	0.000	0.533	0.000
Phonology[b]		0.251	0.000	0.797	0.000
	Case	0.233	0.070	0.890	0.118
	Control	0.406	0.000	0.584	0.000
Morphosyntax[b]		0.181	0.016	0.900	0.010
	Case	0.192	0.200	0.901	0.166
	Control	0.148	0.200	0.957	0.577
Semantic[b]		0.254	0.000	0.862	0.001
	Case	0.257	0.028	0.807	0.011
	Control	0.243	0.009	0.809	0.003
ABR μV^{2} [b]		0.177	0.020	0.805	0.000
	Case	0.288	0.007	0.635	0.000
	Control	0.124	0.200	0.913	0.112
MLR μV^{2} [b]		0.185	0.013	0.848	0.001
	Case	0.159	0.200	0.902	0.169
	Control	0.193	0.092	0.885	0.038

A bivariate statistical analysis, i.e. the analysis of the relationship between being a case/ control and quantitative variables (comparison of means), was also carried out. The results are shown in Table IV. According to these results, being a case is related to lower scores in phonetic, phonological and morphosyntax skills.

Table 4. Bivariate statistical analysis

	Age	Phonetic	Phonologic	Morpho-syntax	Semantic	ABR µV²	MLR µV²
Test U M.-W.	93.5	39.500	19.000	16.000	65.000	78.000	72.000
S. mean (p)	0.706	0.002*	<0.001*	<0.001*	0.076	0.288	0.184

* Statistic significance

Finally, the relationship between electrophysiological findings and linguistic scores (correlations and clouds of points) was analyzed, with the following results. There is a correlation between the phonetic level and the response between 3 and 6.5 milliseconds (ABR), linear regression with prediction interval of the mean at 95%, with a correlation coefficient of (Spearmen rho): -0.407 (p = 0.028) (Figure 4).

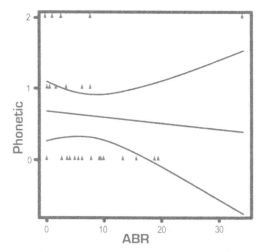

Fig. 4. Correlation between phonetic and ABR.

A correlation between the phonological level of language and ABR was found, with a correlation coefficient of (Spearman rho): -0.320 (p = 0.090) with an interval of 95% (Figure 5). Outliers in figure 4 and figure 5 do not change the meaningful or direction of results, only the coefficient value decreases. The Spearman correlation avoids the extreme values effects. No correlation was found between the morphosyntax level of language and the ABR with a correlation coefficient of (Spearman rho): 0.097 (p = 0.615) (Figure not shown).

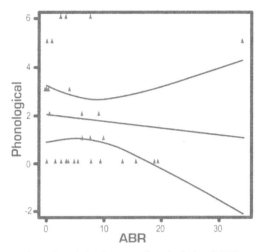

Fig. 5. Correlation between phonological and ABR.

It was observed that there was no correlation between scores in semantic level and ABR, with a correlation coefficient of (Spearman rho): 0.192 (p= 0.318) (Figure not shown). The correlation between scores on the phonetic level and MLR was moderate and statistically significant, with a correlation coefficient of (Spearman rho): -0.487 (p = 0.007) (Figure 6).

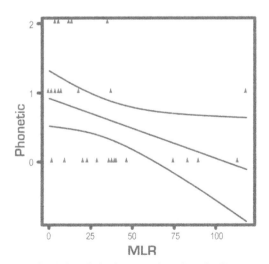

Fig. 6. Correlation between phonetic and MLR.

The correlation between the phonological level of language and the MLR was moderate and statistically significant with a correlation coefficient of (Spearman rho): -0.433 (p =0.019) (Figure 7).

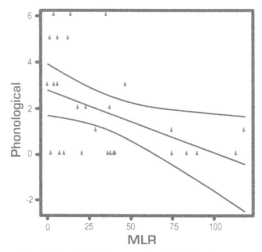

Fig. 7. Correlation between phonological and MLR.

The correlation of morphosyntax level and the MLR were not statistically significant, with a coefficient of correlation (Spearman rho): 0.191 (p = 0.322) (Figure not shown). No correlation was found between the semantic level of language and MLR, with a correlation coefficient of (Spearman rho): 0.045 (p = 0.817) (Figure not shown).

4. Discussion & Conclusions

It has been exposed by other authors that, the persistent character of the OM might lead to auditory processing alterations.[19] This highlight the importance of measuring both the duration and the consequences for each episode of OM suffered in early childhood. The interaction between brain plasticity and sensory deprivation explains the central origin of auditory processing disorders. It is possible that the changes in cortical structures caused by environmental factors, auditory in this case, remain until the next modification occurs. Consequently the fact that the structure has been modified explains the cortical origin of these disorders. The distortion in frequency perception plays a central role in the auditory dynamic process and consequently in the phonological level. The articles reviewed for this study suggest that the areas of language with alterations are the phonological and morphosyntax levels.

The specific language impairment is extremely wide-ranging; and it is characterized by the heterogeneity of sufferers. This is why it is highly difficult to determine a single common cause for this vast range of symptoms; we have to consider socio-educational and genetic factors, amongst others. Therefore, not all children who present a phonological disorder have a history of auditory deprivation in early childhood, but clearly these must to be considered as a risk in presence of other predisposing factors.

According to the results obtained we can make the following conclusions: there is correlation between a history of right otitis media in early childhood and receiving less energy in the auditory cortex in the MLR (between the 15 and 30 milliseconds); there is a correlation between a history of right otitis media in early childhood and lower scores in the phonetic, phonological and morphosyntax levels of language; and there is correlation between lower energy of responses between 15 and 30 milliseconds (MLR) and lower scores on phonetic and phonological levels.

These outcomes support the idea of an ERP marker to diagnose dynamic auditory processing disorders or at least as a risk factor measure. The simplicity of the procedure allows the implementation in children with low levels of attention (Attention Deficit Disorder with Hyperactivity), motivation (Pervasive Developmental Disorders) and/or comprehension (Specific Language Impairment).

A computational model of the sensorimotor basis of phonological acquisition simulates the sounds detection and categorization through spectral properties of utterance. This model of perception specializes in segmenting and categorizing the acoustic signal into discrete phonetic events and matches them with a gesture learned by an articulatory model.[37] Our study follows this line in a clinical application and highlights the potential consequences of the disruption in any stage.

Acknowledgements

We are grateful to all the children and their parents who participated in the study.

References

1. E. Lenneberg, *New Eng J Med.* **266**, 385 (1962).
2. D. Johnson, *J Commun Disord.* **41**, 20 (2008).

3. L. Falkenberg, *Brain Cognition.* **76**, 276 (2011).
4. I. De Witt and J. P. Rauschecker, *Proc Natl Acad Sci U S A.* **109**, E505 (2012)
5. M. V. Popescu and D. B. Polley, *Neuron.* **65**, 718 (2010).
6. F. B. Del Castillo, *Enfer Infec Micro Clin.* **26**, 505 (2008).
7. L. O. Redaelli de Zinis, C. Campovecchi, G. Parrinello and A. R. Antonelli, *Int J Audiol.* **44**, 593 (2005).
8. J. K. Pukander, *Acta Otolaryngol.* **94**, 479 (1982).
9. J. Klein, *Pediatr Infect Dis J.* **3**, 389 (1984).
10. J. W. Gravel, *J Speech Lang Hear R.* **43**, 631 (2000).
11. P. Menyuk, *Ann Oto Rhinol Laryng.* **89**, 257 (1980).
12. M. L. Uhari, *Eur J Pediatr.* **157**, 166 (1998).
13. D. K. Teele, *J Infect Dis.* **162**, 685 (1990).
14. J. W. Gravel, *Acta Otolaryngol.* **115**, 279 (1995).
15. S. F. Friel-Patti, *J Speech Hear Res.* **33**, 188 (1990).
16. P. N. Nelson, *J Acoust Soc Am.* **97**, 1830 (1995).
17. P. K. Stelmachowicz, *J Acoust Soc Am.* **98**, 1388 (1995).
18. R. L. Schwartz, *J Speech Hear Disord.* **50**, 141 (1985).
19. D. Moore, *J Commun Disord.* **40**, 295 (2007).
20. J. W. Hall, J. H. Grose and H. C. Pillsbury, *Arch Otolaryngol Head Neck Surg.* **121**, 847 (1995).
21. J. B. Roberts, *J Speech Hear Res.* **34**, 1158 (1991)
22. J. L. Paradise, C. A. Dollaghan, T. F. Campbell, H. M. Feldman, D. K. Bernard, D. K. Colborn, H. E. Rockette, J. E. Janosky, D. L. Pitcairn, D. L. Sabo, M. Kurs-Lasky, C. G. Smith, *Pediatrics.* **105**, 1119 (2000).
23. M. S. Rovers, H. Straatman, K. Ingels, G. J. van der Wilt, P. van den Broek and G. A. Zielhuis, *Pediatrics.* **106**, E42 (2000).
24. H. Pillsbury, J. Grose and J. W. Hall, *Arch Otolaryngol Head Neck Surg.* **117**, 718 (1991).
25. D. L. Bishop, *Speech and language impairments in children: causes, characteristics, intervention and outcome.* Psychology Press, Oxford. (2001)
26. L. Bosch, *Rev de Logo, Fon y Audi.* **7**, 195 (1987).
27. G. Aguado, *El desarrollo de la morfosintaxis en el niño. Manual de evaluación del TSA.* CEPE, Madrid. (2002).
28. L. D. Dunn, PPVT-III. *Peabody, test de vocabulario en imágenes.* TEA, Madrid. (2006).
29. A. L. Berg, B. A. Prieve, Y. C. Serpanos and M. A., *Pediatrics.* **127**, 269 (2011).
30. M. M. Báez-Martín and I. Cabrera-Abreu, *Rev neurol.* **37**, 579 (2003).
31. P. Ucles, *Dyslexia.* **15**, 72 (2009).
32. G. K. Yule, *An Introduction to the Theory of Statistics.* Charles Griffin & Co, London. (1969).
33. H. W. Mann, *Ann. Math. Statist.* **18**, 50 (1947).

34. S. W. Shapiro, *Journal of the American Statistical Association*, 1343 (1968).
35. W. David, *Weighing the Odds: a Course in Probability and Statistics*. University Press, Cambridge. (2001).
36. J. H. Roberts, *J Dev Behav Pediatr.* **25**, 110 (2004).
37. K. L. Merkey, *The sensorimotor foundation of phonology; A computational model of early chilhood articulatory development*. PhD thesis, University of Colorado, Boulder-Colorado. (1994).

High-Level Cognition

THE INFLUENCE OF IMPLEMENTATION ON "HUB" MODELS OF SEMANTIC COGNITION

OLIVIA GUEST, RICHARD P. COOPER and EDDY J. DAVELAAR

Department of Psychological Sciences, Birkbeck, University of London, Malet Street, London, WC1E 7HX, United Kingdom

Rogers et al.[1] present a model of semantic cognition – the "hub" model – that reproduces the behaviour of neurologically healthy and neurologically impaired individuals on a range of tests of semantic knowledge. The model and associated theory provide a comprehensive explanation for deficits, such as semantic dementia, by appealing to the breakdown of attractors within a recurrently connected system following damage. We report findings from an attempted replication of the Rogers et al. model. While normal behaviour was reproduced, lesioning the reimplementation did not fully replicate the behaviour of the original model, meaning that the reimplementations contain healthy semantic representations which are in line with the hub theory, but the effects of damage on the structure of the semantic representations are not theoretically accounted for. The hub theory predicts that after damage semantic representations must decay in certain ways in order to give rise to patient behaviour. Our results show that the reimplementations do not fully exhibit these symptoms. This suggests that while semantic impairments reminiscent of patients may arise following lesioning of the hub model, such patterns are not a necessary consequence of the model as initially described. We discuss the implication of this apparently negative result for the hub theory of semantic cognition, focusing on differences between our reimplementation and the implementation of Rogers et al., and on the theory-model relationship more generally.

1. Introduction

Semantic cognition comprises a set of cognitive processes that give percepts meaning, allowing for the formation of relations over both concepts and percepts. Various semantic tasks have been developed that test a subjects' ability to access and use semantic concepts, given percepts. These tasks are designed to be administered to both neurologically healthy individuals and patients with semantic deficits (e.g., semantic dementia, herpes simplex virus encephalitis, semantic aphasia, etc.) to determine the exact nature of patient deficits. Patient and healthy participant data may further be used to inform and benchmark cognitive modelling efforts, which aim to account for the ability to perform semantic cognition tasks and predict the breakdown of semantic memory.

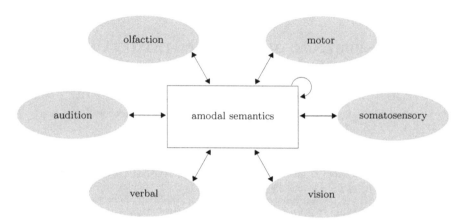

Fig. 1. An overview of the semantic hub and its modal spokes, based on figure 2 in Lambon Ralph et al.[2].

One influential account of semantic cognition is the "hub" theory of Rogers et al.[1]. According to this theory, modality-specific perceptual inputs (e.g., visual, aural, motor, somatosensory, etc.) are reciprocally connected to a central amodal hub, as shown in 1. The information passed between the hub and its spokes allows for retrieval of semantic associations (e.g., visualising a dog based on hearing a bark), identification (e.g., calling a picture of a dog "dog"), categorisation (e.g., classifying a poodle as a "dog", "mammal" and "animal"), and generation (e.g., describing, drawing, or imitating a "dog"). Damage to the connectivity within the amodal semantic hub, and between the hub and the modal spokes, is proposed to give rise to the deficits seen in patients. The model thus aims to provide an explanation for both normal and impaired semantic cognition.

Rogers et al.[1] present a connectionist implementation of the hub theory. Their model, shown in 2.1, consists of a subset of the possible modalities, allowing for visual, verbal and name input/output, and is specifically designed to account for the effects of neurodegeneration as seen in semantic dementia (SD) patients. When undamaged, the Rogers et al. model performs at ceiling on four tests of semantic cognition, as do neurologically healthy participants. However, after removing a random subset of all connections (by setting the corresponding weights to zero), the model shows the same qualitative patterns as SD patients. Thus, the model is held to capture the complexity of semantic cognition required to explain both normals and SD patients.

The mechanism that underpins concepts – both in the hub theory and in the connectionist implementation – is the emergence of attractor states. Such states arise in dynamical systems that have recurrently connected components. Given partial input the system state gravitates towards the centre of a basin of attraction, thus recreating the full multi-modal experience of the concept. The hub theory proposes that, as a result of lesioning connections, neighbouring attractor basins coalesce, creating larger more generalised concepts. Attractors that are proximal in semantic space merge to represent a more general concept, as reflected in SD patients' responses on semantic tasks.

Based on three reimplementations, we explore the claim that the hub theory as described by Rogers et al.[1] is sufficient to yield models with the required attractor dynamics. In the following sections, the effect of implementation differences on model performance on semantic tasks is discussed with the aim of illuminating the relationship between the hub theory and its implementation. We conclude by considering the general relationship between models and theories.

2. Three implementations of the hub theory

2.1. *General architecture*

The hub model, as presented in Rogers et al.[1], is a real-valued recurrent neural network consisting of three pools of input/output units: 40 name units, 64 visual units, and 111 verbal units (further subdivided into 61 perceptual, 32 functional, and 18 encyclopaedic units). Name units represent natural language labels (e.g., "car"), visual units code for visual perceptual features (e.g., "is blue"), and verbal units assume the role of general verbal properties that are perceptual (e.g., "makes noise"), functional (e.g., "can cut"), and encyclopaedic (e.g., "is living"). These units are bidirectionally connected to 64 fully recurrent hidden units, as shown in 2.1. The input/output pools represent the sensory spokes, and the hidden units represent the amodal semantic hub. Activation spreads from one or more spokes to the hub and from the hub back to every input/output pool, thus functioning as a pattern-completing auto-associator.

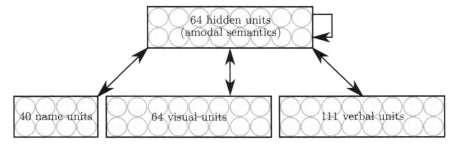

Fig. 2. The hub model's neural network topology, based on figures 1 and 3 in Rogers et al.[1].

2.2. *Pattern set*

Rogers et al. provide a probabilistic template for generating appropriate training sets. We use this template (cf. fig. 3, Rogers et al.[1]) to create a pattern set, equivalent in structure to the original Rogers et al. (as evidenced by hierarchical cluster dendrograms), for training and testing our reimplementations. According to the template, mutually exclusive subsets of visual and verbal features underpin the main distinction between man-made and inanimate objects, as shown in 2.2. Other structural properties are: that the two domains are subdivided into 6 categories (mammals, birds, fruit, vehicles, household objects and tools); that verbal sub-patterns include a single feature present to denote category and domain membership; and that names consist of a single uniquely activated unit, thus creating 40 orthogonal name bit vectors. Some names are shared between certain visual/verbal sub-patterns in order to create category-level names, thus giving rise to archetypal patterns (e.g., labelling an animal "dog" as opposed its breed name).

The elements of the training set are binary vectors each with 215 bits. Each vector has the following bits set: *a*) the individual visual or verbal features it possesses (e.g., "is red", "has legs"); *b*) the localist orthogonal bit vector that constitutes the name sub-pattern (e.g., "robin"); and *c*) the localist category and domain membership units within the verbal sub-pattern (e.g., "is mammal", "is tool"; in 2.2 these are represented by the last 7 units). Based on this structure, we created a pattern set consisting of 48 items that abides by the above constraints. In other words, each pattern consists of a name, which contains no intrinsic information, followed by a set of visual and verbal properties, which contain shared and distinctive features that enable the network models to infer a similarity structure.

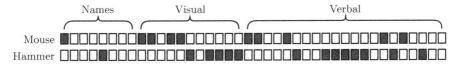

Fig. 3. Two examples of simplified patterns. Solid rectangles represent activated features in the visual and verbal sub-patterns (e.g., "has fur"), while empty ones represent features that are not present.

2.3. Training algorithms

We report three implementations of the hub theory using the architecture and pattern set described above.

2.3.1. BPTT$_1$

The first network was trained using epochwise back propagation through time[3], following the procedure of Rogers et al.[1] where specified. BPTT is a variant of back propagation that involves "unrolling" a multi-layered feedforward version of the recurrent network and training the weights using standard back propagation within this new unrolled network[4]. When the learning phase is completed the network is reverted back to its normal recurrent state. Following Rogers et al., the network was settled for 28 steps during training. As in Rogers et al., the input units were clamped (i.e., forced to take on their target values) for twelve of these steps. We refer to this method of training as BPTT$_1$.

2.3.2. BPTT$_2$

An alternative method, which we refer to as BPTT$_2$, is to clamp the targets to the outputs for the full 28 settling steps – every other aspect of this training procedure is identical to BPTT$_1$. This reduces the noise in the error signal during training resulting in an order of magnitude fewer epochs to learn the training set.

2.3.3. BM

A radically different way of implementing the model is to use a Boltzmann machine (BM). BMs are a type of binary-valued recurrent stochastic neural network. This kind of network is able to conform to the topology required by the hub theory and permits the emergence of attractors[5]. Training involves minimising the difference in unit activations between the network settled with all inputs clamped, known as the plus state, and the network settled on each sub-pattern (e.g., just the verbal features clamped), called the minus states[6].

3. Simulation results

3.1. *Normal behaviour*

Semantic cognition can be assessed using: *a*) confrontation naming, which involves giving an appropriate linguistic label to a line-drawing; *b*) word-to-picture matching, pairing a card with a word on it to one with a drawing of the same animal or artefact; *c*) sorting words and pictures, categorising the aforementioned word and picture cards into two piles for each domain (animate/inanimate objects) and into six categories (mammals, birds, fruit, vehicles, household objects, and tools); and *d*) drawing, copying, and delayed copying, which requires sketching from memory and copying line-drawings either directly or after a time delay.

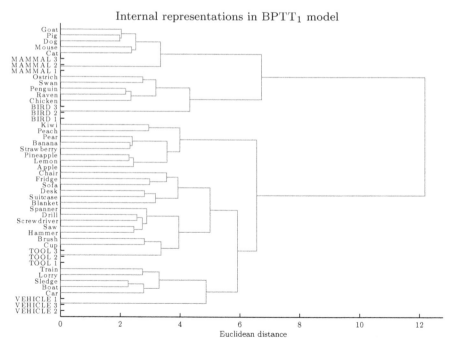

Fig. 4. Dendrogram for BPTT$_1$ internal semantic states sampled 100 times per name sub-pattern input. Category names, shared over three patterns, are capitalised.

Once trained, healthy naming and sorting are possible in all models except the BM. This is due to the inherent stochastic nature of BMs, the need for extra training to better learn the mapping between visual input and name output, and the design of the two tasks. The naming and sorting tasks make use of localist interpretations for names and category membership, which are not part of the BM's learning strategy. This issue can be addressed given greater training, although with the training given in this study the BM does score at ceiling on the word-to-picture matching and the drawing, copying, and delayed copying tasks, which use distributed representations.

As required by the hub theory, all the networks have internal states that allow for the mapping of the perceptual inputs to the output modalities, thus completing each of the four semantic tasks. Fundamentally, the internal semantic space must mirror the categorical and domain structure of the training set. This attractor-space can be represented using a dendrogram as in 4, which shows the Euclidean distance between both individual concepts and between categories and domains. This allows a comparison between the intended categories and those that arise from the structure of the learned attractor states (see figure 5 in Rogers et al.[1]).

Rogers et al.[1] provide a list of qualitative properties that their model's internal representations possess. As shown in 4, our versions also conform to this list. Firstly, the two domains, animals and artefacts, are clearly separated from each other, as are to a lesser extent the six categories. Secondly, the model's representation of category-level names (e.g., "BIRD 1") are classed within their category cluster. And finally, fruit are classed under the domain of inanimate objects, but are in a distinct cluster to the the rest of the artefacts.

3.2. *Damaged behaviour*

Since our reimplementations have healthy internal representations as found in the original hub model, damage can be applied to cause disruptions to the attractor basins. SD-like damage is modelled by setting increasing proportions of all connection weights to zero. This causes the network to be less adept at completing semantic tasks, as propagation of activations both within the hub and between it and its spokes is impaired. Disconnection has a pronounced effect on the semantic attractor landscape; the network can now only manage to represent a subset of the previous 48 concepts, which can be seen in 5. The clusters corresponding to concepts, categories, and domains are now deformed, e.g., the attractors for "cup" and "mouse", from

162

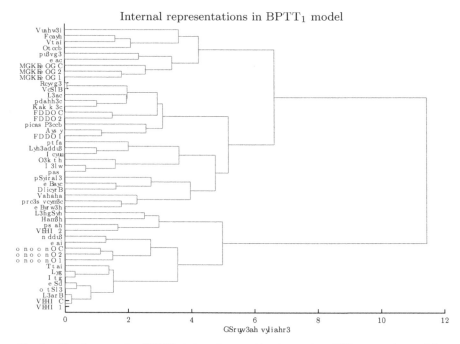

Fig. 5.   Dendrogram for BPTT₁ internal semantic states after 30% connection weights lesioned sampled 100 times per name sub-pattern input.

opposing domains, are now in the same semantic cluster. This merging of conceptual representations from different domains, as opposed to categories within the same domain of knowledge, appears to signal a deviation from the hub theory's requirements.

In the next few sections, we focus on behaviour in the naming and sorting task, as these two tasks are the most problematic in terms of accounting for the patient data in the reimplementations[a]. These two tasks results depend heavily on the network settling to the correct internal state after lesioning damage.

---

[a]Rogers et al.[1] also consider behaviour of the hub model on word-to-picture matching and in the drawing, copying and, delayed copying task. Our reimplementations replicate these results.

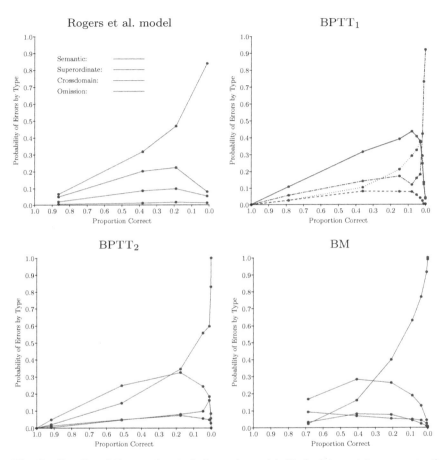

Fig. 6. Results of the naming task for each model. Each data point represents the proportion of error for each type of error at a percentage of connections lesioned: from 0% to 90% in increments of 10% sampled 500 times; in the case of the Rogers et al. results are at 10%, 20%, 25%, and 35% connections lesioned, taken from figure 6 of Rogers et al.

3.2.1. Confrontation naming

The confrontation naming test involves producing an appropriate word when given a visual depiction of an object. For the group of patients this involves a line-drawing and an experimenter to record their response. For the models, following the original task design, naming involves clamping the visual units (representing the input to semantics when looking at a picture) and then allowing the network to cycle for twelve settling steps (cf. p. 217, Rogers et al.[1]). After that the visual units are unclamped and the network is allowed to settle until equilibrium. When the network reaches a stable end-state the name output is inspected and whichever name unit is found to be most active, above a threshold of 0.5, is taken to be the model's response. If no unit has an activation above the threshold, the response is classified as an omission. This method is not applicable to the binary unit states of the BM. Hence, for the BM, the response is derived by finding the pattern closest in Euclidean space to the name output. If more than one name unit is activated the confidence of the BM is taken to be the reciprocal of the number of active name units; any reciprocal under 0.5 is an omission.

There are four important qualitative features of the naming scores found in both the longitudinally tested SD patients (cf. figure 6, Rogers et al.[1]) and in the original model (reproduced in the top left sub-graph of figure 3.2). Firstly, the overwhelming proportion of errors consists of *omissions*, meaning the patient is anomic or the model is unable to activate any name units above the threshold. Omission errors are seen to increase with the progress of neurodegeneration. Secondly, *semantic errors*, which involve confusing an item with another from the same semantic category (e.g., calling a mouse "dog"), initially start off low, then grow to about a quarter of responses, and finally return to a lower proportion. Thirdly, *superordinate errors*, giving a category name to an item that would not be labelled as such by a healthy participant (e.g., calling a dog "mammal"), show the same pattern as semantic errors. Although at all levels of lesioning superordinate errors are lower than semantic errors, reaching only about a tenth of all responses at their highest proportion. Fourthly, *crossdomain errors*, giving an item a name from the opposing semantic domain (e.g., calling a hammer "dog") are extremely rare in both the fifteen[1] patients and in the original model.

The $BPTT_1$, $BPTT_2$, and BM naming graphs in 3.2 show only a partial replication of the naming task scores as discussed above. Firstly, omissions are lower than semantic errors, but in fact they should be consistently above all other errors. Secondly, semantic errors are proportionally the highest error type. Thirdly, superordinate errors are qualitatively a good fit. Fourthly, crossdomain errors occur, when instead they should be at floor levels. This pattern of responses persists even if the value of the threshold, which determines the proportion of responses that are classified as omissions, is varied.

The results of the confrontation naming task run on the three different implementations show that internal representations do not decay in a way that replicates the patients' scores. So while healthy naming is possible, at least within the BPTT reimplementations, the predictions made by Rogers et al.[1] are not met. Specifically, they claim that "[w]ith increasing damage, the model becomes unable to generate any information that individuates items from the same broad domain, and representations within a given domain collapse into a single general attractor from which the model produces only those properties common to the majority of items in the domain. [That is to say, t]he model never names an object with a completely unrelated label, because such names apply only to objects with very distal internal representations" (Rogers et al.[1], p. 218). However, we can see from both the damaged semantic representations $BPTT_1$ has, shown in 5, and from the naming scores in 3.2, that concepts from opposing domains can become much closer to each other than (what should be) neighbouring concepts. This is why a larger proportion of crossdomain errors are produced: attractor dynamics do not necessarily follow the predictions set out by the hub theory.

3.2.2. *Sorting words and pictures*

The sorting task is used to determine the preservation of hierarchical conceptual knowledge in patients. It is carried out by classifying words and pictures into the five categories (Rogers et al. exclude fruit during testing[1]) and into the two domains, respectively named *specific sorting* and *general sorting*. This semantic task is modelled in the same way as the naming task with regards to settling, by clamping the target for twelve settling steps, then removing the target and allowing the model to reach equilibrium (Rogers et al. use this method for all the tasks). Once the network is in a stable state, the verbal units which represent category or domain membership are examined (cf. p. 220, Rogers et al.[1]). For general sorting, the domain unit for animals or for artefacts is used to determine the response of the network, and for specific-level sorting category units are inspected. Whichever unit is most active is taken to be the model's response.

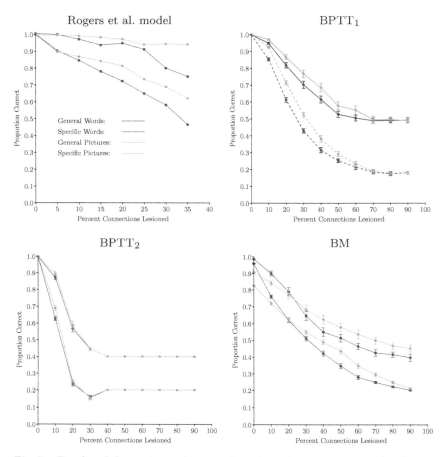

Fig. 7. Results of the sorting task on words and on pictures at a general and specific level of categorisation. Each data point represents the score at the level of lesioning sampled 500 times. Error bars – where present – indicate one standard error about the mean. NB: difference in x-axis scale between the Rogers et al. model and the three reimplementations.

Figure 7 shows the the original model behaviour in the top left corner. This reflects the pattern of the twelve patients tested by Rogers et al.[1], in particular: *a*) the sorting of pictures is more preserved than that of words; *b*) sorting at a general level is retained more so than specific sorting; and *c*) the ability to classify pictures into their respective domains is largely unaffected by lesioning.

The BPTT$_1$ results, shown in the top left sub-graph of 7, indicate that the last of these properties is absent; scores in general picture sorting should be near or at ceiling even after substantial (40%) lesioning. The graphs for the BPTT$_2$ and BM models do not display this property either, but nor do they consistently show the other two qualitative effects. In contrast to the original model, the scores of the three reimplementations for all types and levels of sorting tend towards baseline values (chance for category-level sorting is 0.2 and for domain-level is 0.5 – any slight deviation from these is due to the values of the bias units). Rogers et al.[1] propose that their model of the sorting task is able to follow the patients' scores because "the effect of damage must be quite severe before the system begins to generate incorrect verbal information about such properties" (Rogers et al.[1], p. 218).

The BPTT$_1$ reimplementation manages to show a partial replication, however the BPTT$_2$ and BM do not reflect any aspect of the patient scores consistently. So while in the original model the "difference in the nature of the mapping between surface form and conceptual representations [...] underpins the difference in performance for word and picture sorting" (Rogers et al.[1], p. 221), this does not hold as strongly for the BPTT$_2$ and BM. In addition, in the original hub model "[a]rbitrary mappings are more vulnerable to damage than are systematic mappings" (ibid, p. 221), meaning that word sorting is more fragile than picture sorting; however, this also does not generalise to all our reimplementations.

It should be noted that qualitatively equivalent naming and sorting scores as those seen in the reimplementations in figures 3.2 and 7 are found over many instances of BPTT$_1$, BPTT$_2$, and BM networks (i.e., the results are not an artifact of one set of trained weights). In addition, the training algorithm of the BPTT networks has been varied between epoch-wise, pattern-wise and sub-pattern-wise (weights updated after each name, verbal, visual sub-pattern) and it has been found to also produce qualitatively equivalent naming and sorting graphs.

4. Discussion

McClelland argues that:

> When a model fails to capture some aspect of human performance, it represents both a challenge and an opportunity. The challenge is to determine just what aspect or aspects of the model are to blame for the failure. Because the model is an exploration of a set of ideas, it is not clear which members of the set are at fault for the

models shortcomings. Model failures also present an opportunity: When a model fails it allows us to focus attention on where we might have been wrong, allowing real progress to arise from further investigations.

McClelland[7], p. 21

The work presented here demonstrates that the implications of the ideas embodied in the hub theory do not necessarily capture some aspect of human performance. In other words, because reimplementations do not show the same pattern of errors, as a consequence of not showing an equivalent decay of attractor basins, our results represent both a challenge and an opportunity for further research. The original hub model and the reimplementations we present here constitute an exploration of a set of ideas, some of these ideas might lead to conclusions or give rise to phenomena that might not have been uncovered a priori. By running different models based on a theory, as McClelland[7] claims, the repercussions of the ideas explicitly and implicitly contained in the theory can be illuminated and explored.

Within the hub theory, attractors break down in ways that are predictable and this breakdown is the phenomenon proposed to account for the semantic impairments documented in patients. However this does not appear to hold for all implementations of the hub theory. Specifically, while our three reimplementations are adept at patient modelling on the word-to-picture and drawing and delayed copying tasks, they do not fare well when reproducing patient scores in the naming and sorting tasks; nor do our models exhibit the required pattern of breakdown in their internal representations. This means that the ideas encapsulated within the hub theory can lead to models that are not fully in line with the higher level aims of the theory, i.e., to explain the effects of the neurodegeneration caused by semantic dementia on the semantic cognitive system.

In the original hub model, Rogers et al.[1] describe the breakdown in performance of the hub model following damage as arising because "small amounts of drift may lead the network into an inappropriate proximal attractor, [thus making the model] produce incorrect responses appropriate to a semantically related object[, meaning that the attractor space is] robust even to relatively large amounts of damage, because the system's internal representations must be severely distorted before they drift out of the region to which such properties apply" (Rogers et al.[1], p. 229). This has been shown not to hold for our reimplementations, as errors have been documented that are not semantic relations of the target response, instead they are from the opposing domain of knowledge. This is not documented in the original model, or the patients. In the reimplementations presented here it occurs even given relatively small amounts of lesioning damage.

Why might our reimplementations, when damaged, fail to reproduce the behaviour reported by Rogers et al.[1]? One possibility is that the pattern of breakdown of attractors as required by the hub theory is not a necessary consequence of a recurrent neural network trained with the structure of the training set. The hub theory assumes that attractors drift apart and merge in certain ways, as a consequence of the underlying recurrent neural network substrate, without *requiring* this at a theoretical level. But this assumption does not always hold. Based on this disparity between models and theory, it appears that the hub theory is underspecified as different implementations behave differently. Therefore, some additional theoretical constraint is required if models that implement the hub theory are to be consistent with the patients' behaviour. In our view this constraint should concern the behaviour of attractors following lesioning.

Acknowledgements

We are grateful to Jay McClelland, Tim Rogers, Matt Lambon Ralph, and Anna Schapiro for clarifying aspects of the original hub model.

References

1. T. Rogers, P. Garrard, J. McClelland, M.A. Lambon Ralph, S. Bozeat, J. Hodges and K. Patterson, *Psychological Review* **111**, 205 (2004).
2. M. A. Lambon Ralph, C. Lowe and T. T. Rogers, *Brain* **130**, 1127 (2007).
3. R. Williams and D. Zipser, *Neural Computation* **1**, 270 (1989).
4. R. Williams and D. Zipser, *Backpropagation: Theory, architectures, and applications* , 433 (1995).
5. G. Hinton and T. Sejnowski, *MIT Press, Cambridge, Mass.* **1**, 282 (1986).
6. D. H. Ackley, G. E. Hinton and T. J. Sejnowski, *Cognitive Science* **9**, 147 (1985).
7. J. McClelland, *Topics in Cognitive Science* **1**, 11 (2009).

HIERARCHICAL STRUCTURE IN PREFRONTAL CORTEX IMPROVES PERFORMANCE AT ABSTRACT TASKS

R. TUKKER* and A. C. VAN ROSSUM

Almende BV, Westerstraat 50,
3016 DJ, Rotterdam, The Netherlands
* E-mail: remco@almende.org

S. L. FRANK

Centre for Language Studies,
Radboud University Nijmegen, The Netherlands

W. F. G. HASELAGER

Donders Institute for Brain, Cognition and Behaviour
Radboud University Nijmegen, The Netherlands

We used an Echo State Network (ESN) to investigate cognitive control tasks that involve temporal and policy abstraction, namely an n-back task and a Wisconsin Card Sorting task. In the task conditions where those abstractions were most relevant, separating the abstraction levels increased ESN performance, while the performance was reduced by such a separation in the conditions where the abstractions were less important. This result gives a possible explanation for the topological mapping of abstraction levels over the prefrontal cortex, because of wiring economy in the brain.

1. Introduction

Over the past years, multiple fMRI and other neuroscience results led to the proposal that the Prefrontal Cortex (PFC) is organized hierarchically, with a topological mapping of abstraction levels over the PFC [1-5]. Anterior regions of the PFC are thought to facilitate higher abstraction levels, while posterior regions would be responsible for lower abstraction levels. Badre and D'Esposito [6] distinguished four abstraction types in those studies: domain generality, relational integration, temporal abstraction and policy abstraction. We will investigate tasks involving the two most studied abstraction types, policy and temporal abstraction, suggesting a reason why they may be mapped topologically on the PFC.

Temporal abstraction simply means that certain information is relevant for a longer duration than other information. A prime example is the n-back test. *Policy abstraction* (also known as context abstraction) occurs when a goal or context representation does not deal with physical reality, but instead only concerns other representations. This is for example the case with Wisconsin Card Sorting (WCS). A particular color, shape or number of items determines the motor response (which is not abstract) in the WCS task; however, policy abstraction is introduced because we need the current context to determine which property of the card is currently relevant for the sorting.

We hypothesize that control tasks that involve policy or temporal abstraction share the requirement of information *preservation between* the abstraction levels and information *integration within* the abstraction levels. This simple hypothesis can explain the topological mapping of abstraction levels, because it is likely that more neural connections are required for integration of information than for keeping information separated. Therefore, the total axon length that is required in the PFC is smallest when information is integrated within each area, and preservation of information is obtained by processing it over different areas. That total axon length is in fact minimized by evolution is known as the wiring economy principle [7-9]. Thus, a confirmation of our hypothesis would provide evidence and a possible explanation for the topological mapping of abstractions in the PFC.

So far, the imaging results led to several theories of abstraction levels in the PFC, of which we will discuss the most significant. One very influential theory is presented in the fMRI study by Koechlin *et al*[10]. Three abstraction levels were investigated by varying the number of possible motor responses, the amount of information given by a context cue and the amount of information given by an instruction cue in the beginning of a block of trials. This resulted in differential activation in dorsal premotor cortex (PMd), caudal lateral PFC (lPFC) and rostral lPFC, respectively. This is evidence for a topological mapping of policy and temporal abstraction. Structural equation modelling was used to support the idea of a "cascade of control", from rostral lPFC that supplies temporal information, via caudal lPFC that supplies context information, to the less abstract representations in PMd.

Several research groups replicated these findings, but all in a slightly different way [2,11,12]. For us, one of the studies by Badre and D'Esposito [2] is most important. They varied the amount of competition on three successive policy abstraction levels and additionally one temporal abstraction level. The fMRI result was that PMd, anterior PMd, the inferior frontal sulcus and the frontopolar cortex (FPC) were respectively activated. This led Badre and D'Esposito to propose that each additional level of abstraction might activate an increasingly anterior region in PFC. While both the abstraction levels and the results are similar to those from Koechlin *et al.* [10], the experimental conditions and the activated regions are not exactly the same [2].

The most recent addition to the collection of theories and fMRI results comes from a study by Reynolds *et al* [3]. The different abstraction levels were compared over five conditions: baseline, low temporal and policy abstraction, high temporal and low policy abstraction, low temporal and high policy abstraction and finally high temporal and high policy abstraction. Because the activation levels in a particular region of interest in dorsolateral PFC (dlPFC) were not completely in line with either Ref. 10 or Ref. 2, Reynolds *et al.* [3] suggest there may not be a topological mapping in the PFC at all. Instead, they propose the "adaptive context maintenance" hypothesis, which states that task-relevant information is maintained in the PFC as long as necessary. We will show that while the adaptive context maintenance hypothesis might be correct, we have a reason not to reject the idea of a topological mapping: it can be beneficial for the execution of abstract tasks, assuming wiring economy.

2. Methods

2.1. *Model*

To test our hypothesis, we will employ an Echo State Network (ESN), somewhat similar to the model by Dominey [13]. ESNs are a relatively new type of recurrent artificial neural network, that addresses problems with training procedures of traditional recurrent neural networks [14]. The reservoir of our ESN is formed by a continuous time recurrent neural network. The activity of node i in the reservoir is updated according to a discrete time approximation of the following equation:

$$\tau_i \dot{a}_i = -a_i + f\left(\sum_j W_{ji} a_j - b_i\right) + u_i \tag{1}$$

With τ the time constants vector, a the activation vector, f the activation function (which is the tanh function in our case), W the sparse connection weight matrix, b the bias vector, and finally, u the external input vector.

The reservoir is initialized in such a way that two clusters can be discerned: a 'top' cluster and a 'bottom' cluster, representing anterior and posterior PFC, respectively. The top cluster gives output to the bottom cluster, but does not receive input from it. The bottom cluster is in turn used to train the output layer (see Fig. 1). Because of this hierarchical structure and the random connectivity in the reservoir, information is better integrated within a cluster and better preserved between the clusters. This makes our model well suited for the investigation of the effect of information separation versus information preservation: either less abstract information is offered to the bottom cluster and more abstract information to the top cluster, or all information is offered to both clusters.

The input to the network is given by setting a pattern of high and low activation values in a number of input nodes. These input nodes are connected to the reservoir through a sparse input connectivity matrix for the duration of the first timestep of each trial. Of course, the input pattern over the input nodes changes from trial to trial. To offer specific information to only part of the network, the input connectivity matrix is initialized to allow connections from this specific input node to one of the two clusters only. This is done without changing the overall connectivity of the input weight matrix, thus without changing the expected total input activity (only the location of this input activity is changed). Apart from enforcing those two clusters in the weight matrices, the connectivity is random.

To obtain output from the network, output nodes are connected to the bottom cluster of the reservoir. Each output node corresponds to one possible response and the output node that shows most activation at the end of the trial defines the response of the model. The fully connected output weight matrix connecting the bottom cluster with the output nodes is trained on the desired response with a simple regression procedure:

$$\vec{w}_i = S^\dagger \vec{d}_i \tag{2}$$

This equation gives $\vec{w}_i$, the weight vector for output node i, from $S^\dagger$, the pseudoinverse of the state collection matrix which contains the activation values of all nodes at the end of each training trial, and $\vec{d}_i$ the vector containing the desired outputs for the node per trial. The desired output is set to an arbitrary positive value if the node should be selected at this trial, and to zero otherwise. The parameter settings for the model are summarized in Table 1. Further details of the implementation are described in Ref. 15, which is available at `https://github.com/RemcoTukker/PFC-ESN/` together with the Matlab implementation of the model.

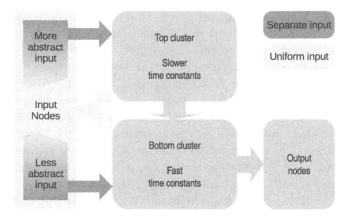

Fig. 1. The architecture of the Echo State Network. In dark gray and light gray the two input methods as determined by the connectivity in the input weight matrix, separated and uniform respectively. In the middle the hierarchical topology of the reservoir fashioned after the PFC. Apart from the constraint on the topology, the connectivity is random. At the right the output nodes for which the weights are trained with a simple regression method. Each output node corresponds to a possible response for a trial in the tasks.

Table 1. Parameter settings of the ESNs.

Property	Value (temporal / policy abstraction)
Number of reservoir nodes	200 / 100
Number of nodes top : bottom	1 : 3
Reservoir connectivity	0.05
Reservoir weight distribution	Uniform, centered at 0 [a]
Top down scaling	1.0 / 0.6
Spectral radius	0.9
Input connectivity	0.1
Input weight distribution	$\mathcal{U}(-2.5, 2.5)$
Bias distribution	$\mathcal{N}(0, 0.3)$
Timesteps per trial	15
Number of training trials	800
Number of test trials	800
Time constants	10 to 10^3 / 10 to 12.5

Note: [a] Top down weights are scaled and bottom-up connections removed. The range of the distribution is determined by the spectral radius and the random connectivity; no fixed values can therefore be given.

2.2. *Tasks*

The first task was designed to investigate policy abstraction and corresponds to the sorting in a WCS task. At each trial, one input node gives abstract information and is set to either low or high activity. This abstract information corresponds to the context in a simple version of WCS: either the cards have to be sorted according to color or according to shape. Depending on this context (or current policy), either the first half of the rest of the input nodes is relevant to determine the desired output, or the second half. These input nodes thus represent the color and shape in WCS, respectively. The input nodes that are not relevant in the current trial are distractors. The task is illustrated in Fig. 2.

The experimental condition in this task is the number of low-level input nodes. If our hypothesis is correct, preserving the information within the abstraction levels is most important with a large number of low-level input nodes, as the abstraction is most important in this situation. On the other hand, integrating information between the abstraction levels is most important in the conditions with a small number of low-level input

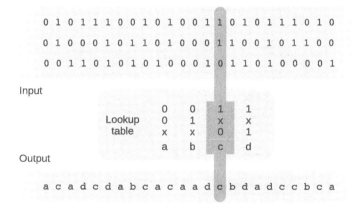

Fig. 2. The version of the policy abstraction task with three input nodes (and thus two low abstraction input nodes). Each column represents one trial. The input pattern at the top is used to set high and low activities in the input nodes at the beginning of each trial. In the lookup table the top row represents the abstract input that is used to determine whether the middle row or bottom row is currently important to find the desired output. The x in the lookup table is a wildcard: the input node can be either 1 or 0, it is irrelevant for the task and thus a distractor. The letters in the output row correspond to the four output nodes in this task, with the highlighted letter corresponding to the output node that should show the highest activation at the end of the current trial.

nodes. Thus, the ESN with topological input should perform better in the conditions with many low-level input nodes, while the ESN with uniform input is expected to perform better in the conditions with a small number of low-level input nodes.

The second task is a generalized version of the n-back test to investigate temporal abstraction. The information presented at one specific, abstract input node has to be retained for n trials, after which it has to be integrated with the current information at the other input nodes to find the desired output. This integration is again dictated by a lookup table, this time with each possible pattern corresponding to one output node. A graphical representation of the task can be found in Fig. 3. To facilitate the retention of information in the reservoir, some of the time constants in the top cluster are randomly increased for this task.

According to our hypothesis, when n and thus the temporal abstraction is small, the most decisive factor in the performance of the network should be the ability to integrate information. Reversely, when n is large, the preservation of the more abstract information should be most important. Thus, we expect the ESN with segregated, topological input to perform better in n-back test when n is large, while the ESN with uniform input should perform better when n is small.

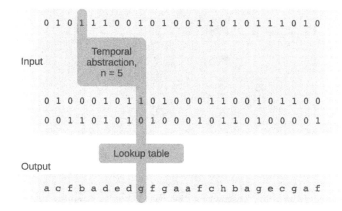

Fig. 3. The implementation of the temporal abstraction task used in the simulations. The basis is a lookup task with three binary input nodes, with each of the eight possible input patterns corresponding to an output node, which is represented by a letter in the output row. Each column in this figure represents a trial and temporal abstraction is thus introduced in the task by presenting the relevant bit from one of the input nodes n trials earlier than the data at the other input nodes, so that this information has to be retained for n trials to find the desired output.

2.3. *Performance Evaluation*

To evaluate the performance of the model on the different tasks, each reservoir was trained on 800 trials and directly afterwards tested on 800 trials, without resetting the network activity at any time. The final performance of each reservoir is simply calculated by dividing the number of correct responses to the test trials by the total number of test trials for the reservoir, yielding a number between zero and one. As each reservoir is random, 200 different reservoirs are initialized and evaluated in this way per condition and per input type, to achieve statistical significance.

3. Results

In Fig. 4 the resulting performance distributions for the temporal abstraction task are shown. In the easiest condition, with memory length $n = 0$, the task is purely the lookup of the current input pattern and both the ESN with uniform input and the ESN with topological input perform perfectly; hence, there is no difference between the two input patterns. Likewise, in the hardest conditions both networks perform weakly and the difference between the ESNs diminishes. However, between those two extremes, we can see that the performance distributions of topological ESNs and uniform ESNs show differences. The same pattern can be observed in Fig. 6, which is a plot of the performance distributions for the policy abstraction task. In both tasks, the difference between the uniform and topological input conditions is highly significant in most cases thanks to the large number of evaluated reservoirs, with $p < 0.001$ in a Mann Whitney U test, as specified in Fig. 4 and Fig. 6.

To simplify the interpretation of these graphs, we choose the difference between the medians of the ESNs with topological input and the ESN with uniform input as a measure of performance difference. In Fig. 5 we can find the differences for the temporal abstraction task. For short memory lengths, the uniformly processing ESNs have an advantage, while the topologically processing ESNs perform better for long memory lengths. The differences of the medians on the policy abstraction task are shown in Fig. 7. Over the conditions in which our performance measure gives a useful result, uniformly processing ESNs have an advantage when a small number of low-level features has to be integrated, while topologically processing ESNs have an advantage when a large number of features has to be integrated.

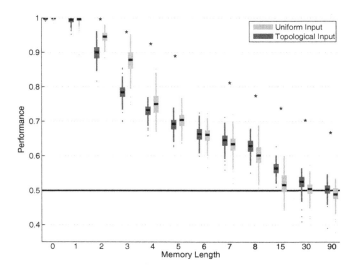

Fig. 4. The distributions of the performance of 200 random ESNs in each condition of the temporal abstraction task. Each box extends from the lower quartile to the upper quartile and the whiskers extend to 1.5 times the interquartile range. The gray line is the performance level that can be reached when the temporal abstract information is not taken into account. Asterisks denote a significant difference ($p < 0.001$) between the ESN with uniform input and the ESN with topological input.

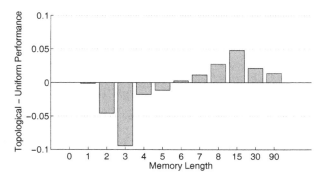

Fig. 5. The difference in median performance between a uniformly processing ESN and a topologically processing ESN as a function of the memory length in a n-back test. The number is positive if the topologically processing ESN performs better.

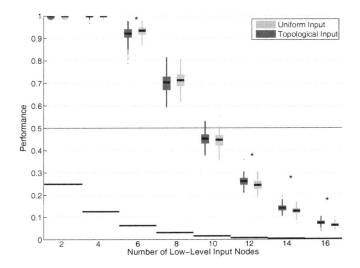

Fig. 6. The distributions of the performance of 200 random ESNs in each condition of the policy abstraction task. Each box extends from the lower quartile to the upper quartile and the whiskers extend to 1.5 times the interquartile range. The dark gray lines are at chance level for each of the conditions, while the light gray line is at the performance level that can be reached by integrating all low-level information but ignoring the high-level input node. Asterisks denote a significant difference ($p < 0.001$) between the ESN with uniform input and the ESN with topological input.

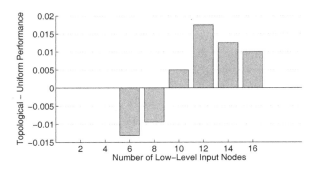

Fig. 7. The difference in median performances between topologically processing ESNs and uniformly processing ESNs at the policy abstraction task. The number is positive if the topologically processing ESN performs better.

4. Discussion

The results are exactly in line with what we would expect from the hypothesis that abstract cognitive control tasks require information to be integrated within an abstraction level and preserved between abstraction levels, as detailed in the methods section. This is in agreement with earlier results: Botvinick has already shown that a topological map of temporal abstraction can be the result of the backpropagation through time learning algorithm in a hierarchical network with an abstract task [4]. Also, such a structure has been shown to partly self-organize using an evolutionary algorithm [16]. The results for the mapping of policy abstraction are for example in line with research by Yamashita and Tani [17], who found self-organized policy abstraction over motor primitives. However, none of these investigations directly addressed the cause of this structure.

Having found requirements for tasks that can influence the way the PFC processes abstraction levels because of wiring economy, we can use these results for the interpretation of the imaging experiments that we described earlier. The work by Koechlin [10] is in agreement with our results, because his experiments included a large difference in timescales: the context had to be remembered for mere seconds, while the episodic control signal had to be remembered for a much longer time, in the order of minutes. This seems reasonable for tasks in daily life as well. The work by Badre and D'Esposito [2] is partly in agreement: a hierarchical mapping of policy abstraction was shown, but the task that included policy abstraction had only a small number of low-level input nodes; a situation in which the uniform processing ESNs performed better in our research. This indicates that the task requirements are not the only factor of importance.

The most recent addition to the debate was made by Reynolds et al.[3], who argue that they did not find any evidence for topological mapping of abstraction and propose the "adaptive context maintenance" hypothesis. This is in conflict with our results: while Ref. 3 suggests an absence of topological mapping of abstractions, our results give a theoretical reason to expect such a mapping, regardless of how the information in the PFC is exactly maintained. Also note that the task used in Ref. 3 was slightly different from the other mentioned studies and that the analysis of the results was done using a rather restricted region of interest. Therefore, we suggest that while Reynolds et al. make a good point for the "adaptive context maintenance" hypothesis, their claim that abstraction levels are not mapped topologically over the PFC goes too far.

Some important limitations of our study have to be discussed as well. First of all, our results may be dependent on our parameter settings. Thus, under different parameter settings the observed effect may not be present or may be unimportant compared to other mechanisms. To mitigate this problem we attempted to keep the number of critical parameters to a minimum and investigated a range of reasonable parameter settings. Unfortunately, an exhaustive gridsearch was not feasible. While we never observed the opposite effect, we could also not confirm the described relation under all parameter settings: especially changing the top-down connection strength led to results in which either the uniform processing ESN or the topological processing ESN always performed better.

Secondly, and arguably more importantly, the structure and organization of the PFC is of course not solely determined by the requirements of exerting control. The two most important factors that have been ignored in this study are the learning of new tasks and the evolutionary history of PFC. Learning new tasks is often considered to be updating or creation of task sets [18]. This is one of the hard tasks the PFC has to solve, making an influence on the structure of the PFC likely. Reynolds and O'Reilly [19] already started exploring this direction and suggested that a hierarchical PFC makes learning easier, by supporting generalization of knowledge or task sets to new domains. Unfortunately, learning is out of the scope of our study because ESNs probably learn very differently from PFC. Regarding the evolutionary history, the model by Hazy, Frank and O'Reilly [20,21] seems to be on the right track already: they explicitly propose the PFC to be an evolutionary extension to the Basal Ganglia.

Finally, it is important to note that we are far from certain how the PFC exactly exerts cognitive control. We assumed that the tasks for the ESNs correspond with the task that the PFC has to solve and concluded that for these tasks, information integration and separation was relevant. However, an important difference is that while we used symbolic input nodes, the PFC might not. This may affect the results, especially because the number of relevant input nodes is one of the identified critical factors. Neural representation remains an active research topic and as yet there are few conclusive answers. Additionally, the natural tasks that have shaped the PFC through evolution are very different from the artificial tasks used in imaging studies and in this modelling work. Thus, we have to be careful when making claims about the causes of structure and organization in the PFC.

5. Conclusion and Future Research

In this article, we argued that the requirements of information integration and information preservation for cognitive control tasks can explain the topological mapping of policy and temporal abstraction levels over the prefrontal cortex. We supported this hypothesis with results from Echo State Networks: in abstract control tasks that require mainly preservation of information, topological processing was advantageous, while uniform processing benefited less abstract tasks that require mainly integration of information. This leads to a mapping over the PFC because of wiring economy: integration of information requires high connectivity while preservation of information requires less connectivity, leading to clustering at different abstraction levels. This result can help us to interpret neuroimaging studies: based on our results, we can expect a functional gradient over the PFC with temporal and policy abstraction levels taking similar roles. This contradicts the recent proposal in Ref. 3. However, we also mentioned some important limitations of our study. To address these limitations and to gain a better understanding of the PFC, we would like to suggest a number of avenues for future research.

First of all, it is important that the imaging results that have been obtained so far are replicated and extended. This could resolve much of the ambiguity of the different conclusions drawn in Refs. 2, 10 and 3 and the present article. Special attention should be paid to the details of the tasks and paradigm: most imaging studies so far mixed different types of abstraction, making a direct comparison with other studies difficult [6]. Moreover, the way in which participants solve tasks should be controlled, because in some cases it is possible to solve an abstract task without using abstraction or *vice versa*. Finally, it may be possible to quantify the amount of control that is necessary using the information-theoretic approaches of either Koechlin and Summerfield [22] or Tononi and Sporns [23].

A second way forward is to integrate aforementioned imaging studies and our own work in a broader context. For example, recently it was found that healthy people show a hierarchical organization of the multimodal network (which includes DLPFC and other prefrontal regions), while this hierarchy was reduced in people with schizophrenia [24]. Interestingly, schizophrenia also compromises the performance on certain cognitive control tasks. Particularly, the masking paradigm (in which a very short stimulus is preceded or succeeded by a mask) shows that patients have a deficit in top-down control of visual information processing [25]. Other diseases, like Alzheimer's disease, may be investigated in a similar way [26]. Thus, it may be possible to link control deficits with changes in PFC organization by using an ESN-based model. Recent developments in physiological studies and the publication of connectomes may help these efforts [27].

Finally, interesting and relevant connections can be made with developmental neuroscience [28] and evolutionary biology [29]. As the infant PFC takes quite some time to develop and achieve adult level of cognitive control, this gives good opportunities to investigate the construction of the hierarchy in the PFC. On a longer timescale, we can study the conditions in which a functional gradient in a hierarchical network will evolve using an evolutionary algorithm [30,16]. Of particular interest in this line of research is the proposal that different types of abstraction may co-occur in nature [4].

Acknowledgments

This work was supported by the European Project REPLICATOR within the work program "Cognitive Systems, Interaction, Robotics" under grant agreement number 216240.

References

1. E. Koechlin and A. Hyafil, *Science* **318**, p. 594 (2007).
2. D. Badre and M. D'Esposito, *Journal of Cognitive Neuroscience* **19**, 2082 (2007).
3. J. Reynolds, R. O'Reilly, J. Cohen and T. Braver, *PloS One* **7**, p. e30284 (2012).
4. M. Botvinick, *Philosophical Transactions of the Royal Society B: Biological Sciences* **362**, p. 1615 (2007).
5. R. O'Reilly, *Trends in Neurosciences* **33**, 355 (2010).
6. D. Badre and M. D'Esposito, *Nature Reviews Neuroscience* **10**, 659 (2009).
7. J. Allman, *Evolving Brains* (W. H. Freeman, New York, 2000).
8. C. Cherniak, *The Journal of Neuroscience* **14**, 2418 (1994).
9. O. Sporns, *Networks of the Brain* (The MIT Press, 2010).
10. E. Koechlin, C. Ody and F. Kouneiher, *Science* **302**, p. 1181 (2003).
11. K. Christoff, K. Keramatian, A. Gordon, R. Smith and B. Mädler, *Brain Research* **1286**, 94 (2009).
12. D. Krawczyk, M. Michelle McClelland and C. Donovan, *Cortex* **47**, 588 (2011).
13. P. Dominey, *Biological Cybernetics* **73**, 265 (1995).
14. H. Jaeger, *The "Echo State" Approach to Analysing and Training Recurrent Neural Networks — with an Erratum Note*, tech. rep., German National Research Center for Information Technology (2001).
15. R. Tukker, Echo state networks for hierarchical cognitive control, Master's thesis, Donders Institute, Radboud University Nijmegen (2012).

16. R. Paine and J. Tani, *Adaptive Behavior* **13**, p. 211 (2005).
17. Y. Yamashita and J. Tani, *PLoS Computational Biology* **4**, p. e1000220 (2008).
18. K. Sakai, *Annu. Rev. Neurosci.* **31**, 219 (2008).
19. J. Reynolds and R. O'Reilly, *Cognition* **113**, 281 (2009).
20. R. O'Reilly and M. Frank, *Neural Computation* **18**, 283 (2006).
21. T. Hazy, M. Frank and R. O'Reilly, *Philosophical Transactions of the Royal Society B: Biological Sciences* **362**, p. 1601 (2007).
22. E. Koechlin and C. Summerfield, *Trends in Cognitive Sciences* **11**, 229 (2007).
23. G. Tononi and O. Sporns, *BMC Neuroscience* **4**, p. 31 (2003).
24. D. Bassett, E. Bullmore, B. Verchinski, V. Mattay, D. Weinberger and A. Meyer-Lindenberg, *The Journal of Neuroscience* **28**, 9239 (2008).
25. C. Gilbert and M. Sigman, *Neuron* **54**, 677 (2007).
26. O. Sporns, *Frontiers in Computational Neuroscience* **5** (2011).
27. O. Sporns, *Annals of the New York Academy of Sciences* **1224**, 109 (2011).
28. K. Supekar, M. Musen and V. Menon, *PLoS Biology* **7**, p. e1000157 (2009).
29. R. Dunbar and S. Shultz, *Philosophical Transactions of the Royal Society B: Biological Sciences* **362**, 649 (2007).
30. J. Bullinaria, *The Importance of Neurophysiological Constraints for Modelling the Emergence of Modularity*, in *Computational Modelling in Behavioural Neuroscience: Closing the Gap Between Neurophysiology and Behaviour*, eds. D. Heinke and E. Mavritsaki (Psychology Press, 2009), pp. 187–208.

INTERACTIVE ACTIVATION NETWORKS FOR MODELLING PROBLEM SOLVING

P. MONAGHAN

Department of Psychology, Lancaster University
Lancaster, UK

T. ORMEROD

Department of Psychology, Lancaster University
Lancaster, UK

U.N. SIO

Department of Psychology, Carnegie Mellon University
Pittsburgh, USA

Interactive activation networks have a long pedigree in the cognitive sciences, and have been widely used to illustrate phenomena of conceptual processing and memory. However, despite numerous proposals regarding their relevance, there are few implementations of interactive activation networks for simulating cognitive processing in problem solving tasks. This chapter discusses potential obstacles to applying such models to problem solving, accounting for this historical lack, and points to ways in which actual implementations can circumvent these obstacles and lead to insights into not only problem solving but also properties of semantic memory more generally.

1. Interactive activation networks in the cognitive sciences

Interactive activation models have been widely used in the cognitive sciences to simulate cognitive processing. McClelland[9] generated an interactive activation network to simulate several features of semantic memory, using very simple properties of interacting units. In the first implementation, the jets and sharks model, units in the model represented individuals or properties of individuals in the musical *West Side Story*. Excitatory connections linked units representing properties of individuals to and from those individuals. The model was successful in reflecting several features of human memory, such as retrieval of knowledge from partial information, and spontaneous generalization on the basis of interacting knowledge resulting in prototype formation and stereotyping. The fact that such complex, interactive effects could emerge from a simple architecture paved the way for low-level, highly-constrained models to demonstrate the cognitive processes underlying apparently complex human behaviour.

Interactive activation networks were, in parallel, increasingly used to reflect detailed aspects of semantic memory. After Quillian's[15] first attempts to simulate effects of semantic similarity in interactive activation models (his wonderfully named "understanding machine"), Collins and Loftus[4] provided a more detailed specification of the properties of semantic memory to be implemented, in particular the greater activation that passes between closely associated words in memory. Raaijmakers and Shiffrin[16] later provided a full implementation of an associative memory network, involving both spreading activation between units and decay of activation for individual units to simulate an array of short-term memory effects.

Interactive activation networks also have a long association with *theoretical* descriptions of problem solving processing. Mednick[10] proposed the process of "spreading activation" in an interactive activation network as a means to describe going from the problem to the solution for remote associates tests. Remote associates tests are problems where the participant is given three words, and is required to find a word that connects all three, for instance, *broken, clear,* and *eye,* or, as another example, *high, book,* and *sour.* [We provide the answers to these items in the *Conclusion,* in order to leave time for the reader to incubate the problem sufficiently for an increased likelihood to solution discovery.] However, the application of such models to problem solving has proceeded with rarity. Before we summarise computational approaches to problem solving using interactive activation networks, and our own implementation of a model of RAT problem solving, the next section describes several reasons why their implementation has been so limited in problem solving research.

2. Obstacles for interactive activation models of problem solving

One source of inhibition in attempts to develop models of problem solving is that the phenomena to be simulated are either too complex, too diverse, or at different levels of description which renders lower-level interactive effects inappropriate for an adequate model (see Russel and Norvig[17] for a review). An attempt to resolve the apparent tension between high- and low-level processing models has been conducted by Helie and Sun.[7] They describe a large-scale computational model that spans both strategic, explicit problem solving processes and implicit, interactive activation effects. Their model exemplifies dual-system theories of reasoning[6], but describes specific points at which the systems may interact and co-influence one another. Helie and Sun show the significance of implicit effects of memory and knowledge representation in problem solving, thus, models such as interactive activation networks may have a valuable role in accounting for high-level problem solving phenomena.

Even if the high- versus low-level differences can be resolved, another of the principal difficulties facing the implementation of problem solving as a consequence of spreading activation in interactive activation networks is time course. Problem solving is typically slow, whereas spreading activation is traditionally described in terms of times orders of magnitude quicker. For instance, in remote associates test performance, participants are given between 5s and 60s to solve problems, and frequently require much of this time to discover an answer.[1] Effects of incubation are even more protracted, with participants typically leaving problems aside for at least a few minutes[19], hours[2,21] or, in some cases, up to a day.[20] In contrast, spreading activation between neurons requires a few milliseconds. How, then, can the metaphor of spreading activation for problem solving be anything more than that – a metaphor?

One possible response is that spreading activation among complex concepts, such as those required for problem solving, requires neural assemblies to be activated, rather than single neurons, and spreading activation among complex neuronal representations may require a longer time-scale than action potentials transmitting between individual neurons (see Sio, Monaghan and Ormerod[20] for discussion of this point). Yet, the time-scale for semantic concepts appears still to be a shorter time-scale than the spreading activation described in problem solving. For instance, semantic priming effects typically last in the order of a few hundred milliseconds[3], and activation in semantic networks representing short-term memory typically decays within a few seconds.[16] But for longer-term incubation effects, it is feasible that very low levels of activation remain within the semantic network engaged for a particular problem, that may later influence problem solving performance after long incubation periods[2], even though the effects are below detection levels for observation in semantic priming studies.

Related to this difficulty of time-scale is the issue of open goals, or Zeigarnik effects[24] in problem solving. This is the observation that memory for undischarged tasks is better than for tasks that have been resolved. Zeigarnik's original observation was for waiters' remembering diners' orders, but the same process has been used to describe effects of unsolved problems, whereby information from unresolved problems can remain active in memory. It is also related to incubation effects in problem solving, whereby leaving an unsolved problem aside for a period of time can improve performance when the problem is returned to at a second attempt.[21] Previous models of problem solving have strived to explain how Zeigarnik effects may be simulated in computational models, converting a description of a phenomenon into an explanation. The next section describes two such attempts.

3. Computational implementations of problem solving

Moss[11] partially implemented an interactive activation network of Smith's[22] algorithm for remote associates test problems. The principal aim of the approach was to demonstrate the ineffectiveness of a model that does not specifically implement effects of open goals on problem solving performance, thus showing that, without specific implementation of Zeigarnik effects, an interactive activation network will be inadequate to describe problem solving performance.

In Moss' model, one of the stimulus words from the test problem was selected, and an associate of this stimulus word was generated. Then, a second stimulus word was selected and the associated word was assessed for whether it was an associate of the second stimulus word. If the association was successful, then the third stimulus word was determined for whether it was also an associate. If at any stage, the associated word did not match as an associate for all words currently selected, then the algorithm returned to the first step and an alternative associate was generated, until a match was found. In Moss' model, an incorrectly selected word was marked so as to prevent repeated selection, but there was a decay on these marks, so that over time, the avoidance of repetition would decline.

Moss' model was able to simulate remote associate test performance for a small set of problems by adjusting parameters of the interactive activation network, but, using these parameters, was not able to closely simulate data from problems where participants were given a hint to a previously unsolved problem. In the model, a hint was implemented by increasing the activation of the hint word in the network temporarily. Behavioural performance showed that a hint results in improved problem solving compared to no hint for previously unsolved problems (see also Smith and Blankenship[19]). However, the model did not predict that performance with a hint would be better than without a hint for previously unsolved problems. Moss interprets this as showing that an additional mechanism for representing open goals, or unsolved problems, in the model was required.

It is possible, however, that open goals may not be required to be explicitly represented in an interactive activation network in order to predict effects of hints on problem solving. The model's performance for unsolved problems with or without hints was identical, and this could be due to an imprecise representation of the activation of hints within the network, or due to too fast decay. In essence, there was no effect of hints on problem solving in the model, but this may be due to the particular dynamics of the model, rather than a need to hold open unsolved problems in the representation of the model. For instance, residual activation in an interactive activation model could be maintained in order to manifest the interactive effect of an activation of a hint word, close to the target word for the remote associates test item.

Accounting for open goals, or the Zeigarnik effect, was also an important motivation for Ohlsson's[13] research, who considered the impact of long incubation effects on problem solving. If spreading activation in networks of words is fast, as predicted by dynamics of biological neural networks or, cognitively, by models of short term memory (e.g., Raaijmakers and Shiffrin[16]) then the fact that incubation effects could be observed over long periods of time for unsolved problems required explanation that went beyond patterns of activation in interactive activation networks. Consider the effect of multiple problems presented to an interactive activation network. For a single problem, a pattern of activation would be observed quickly spreading through the network. Then, for the next problem, another pattern of activation would be observed across a different set of units in the network, and so on for additional problems. In a standard problem solving experiment, multiple problems are presented, which would, if activation was maintained over long periods, create an explosion of overlapping activations in a broad network. With a decision algorithm, such as that used by Moss[11], this would entail that very many potential target answers from a range of problems would be activated for each problem when re-presented after a period of incubation.

To address this difficulty, Ohlsson[14] suggested a mechanism whereby an associative network would adjust its architecture as a consequence of attempted but failed attempts at a solution. If a link between information was attempted but resulted in a judgment by the decision maker (either implicit or explicit) that this route did not advance the problem solver towards the solution, then that route would be pruned from the representation so that alternative avenues for solution search could be explored. Such an adaptation has the advantage for simulating Zeigarnik effects in that later return to the problem confronts the problem solver with an already adapted network, meaning that more effective solutions can be discovered, reflecting incubation effects in problem solving. Once a problem has been effectively solved, then the adaptive pruning of the network can be withdrawn and the full network available for future problem representations.

3.1. *Addressing the obstacles*

The difficulties for an implemented model of problem solving, raised above, could be overcome by subtle adaptation of parameters, or via additional parameters to mark open goals in the problem-solving network, or alternatively, additional parameters to indicate restructured associative routes through a network. Instead of imposing additional assumptions on a model, an alternative solution is to acknowledge the actual structure of semantic memory. In doing so, we provide a model that can account for a larger range of phenomena than models incorporating ad hoc parameters.

Associative networks in semantic memory have been shown to be sparse in terms of interconnections, and have a small world structure.[23] That is, each node in a network is associated to a small set of other words, and these words, in turn, tend to be interconnected to one another within densely interconnected clusters. Words within these clusters are then sparsely connected to some words that occupy other interconnected clusters of words. In terms of an interactive activation network, this has many consequences, not least that multiple remote associates problems could potentially be represented simultaneously in a network with little or no interference between them, because they activate distinct small worlds within the overall semantic network of associates. Thus, open goals could be represented by trace activations of units from previously presented problems, which are then reinvigorated when the initial problem is presented again. So, this small world structure obviates the problem of overlapping problem representation from activations remaining within the semantic network from unsolved problems, without necessarily requiring a separate open goal mechanism or restructuring of the semantic network.

Schilling[18] described such a small world structure in a theoretical model of creative problem solving. She suggested that, once a local set of associates had been exhausted by the problem solver, then movement to other small worlds, representing more distant associates to the initially presented problem, could be activated for search. She viewed this move between small worlds as an insight mechanism, yet it is not incompatible with a search algorithm similar to that used by Moss[11] and proposed by Smith[22] whereby the most highly activated unit serves as a possible solution. Once the local small world network has been exhausted, by selecting then suppressing the highest activated near associate, then associated words that are more distant may become the most highly activated unit in the network.

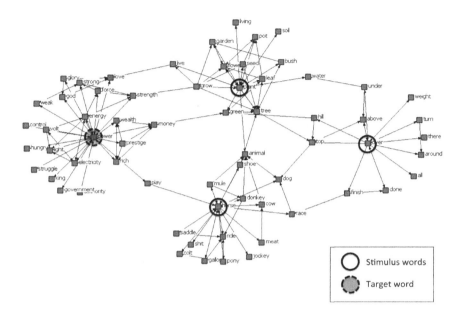

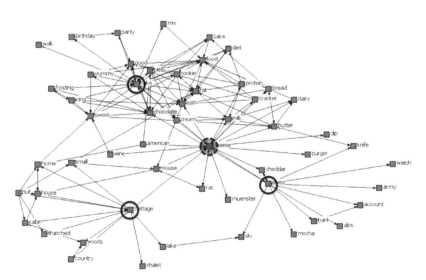

Fig. 1. Two interactive activation networks for an easy and a hard problem: *plant, over, horse* (top), and *cottage, swiss, cake* (bottom).

Another method, not inconsistent with this approach, is to adopt a "foraging" algorithm, where local associates are explored as possible solutions to the problem, until the yield or return of the local search diminishes to a point where it is unrewarding, then a move to another region of the semantic network for search may proceed. Such a model has been implemented by Hills, Jones and Todd[8] to simulate behavioural studies of free association generation (where participants are given a word and required to generate all words that are associated with that word).

In the next section we present an interactive activation model of remote associates problem solving that provides a proof of concept for the small world structure of semantic memory addressing many of the issues that have proven obstacles to modelling problem solving.

4. An interactive activation model of problem solving

We implemented an interactive activation model of semantic memory in order to simulate problem solving performance in remote associates tests. For each remote associate test, a different network was generated. The nodes in the network were all words with associates to and from the stimulus and target words in the remote associate test, as reported in free association norms.[12] Weights between nodes were the strength of the forward or backward associations between words as reported in the Nelson et al.[12] database. Figure 1 shows examples of two fragments of the interactive activation network related to each of the two remote associates problems presented in section 1.2.

Activation passed between units in the network as a linear function of the sum of activation of connecting units multiplied by the strength of the connection. For unit j in the network, the activation of the unit, a_j, is given by Equation 1.1, where a_i is the activation of unit i connecting to unit j and w_{ij} is the strength of the association from unit i to unit j. The activation of units within the network decayed across time (indicated in Equation 1.1 as $-d$).

$$a_j = -d + 0.1\sum_i w_{ij} a_i \qquad (1.1)$$

To initiate activation in the network, at the first time step, the activation of the stimulus words for the problem were clamped at 1. Activation cycled in the model for 50 time steps, and at each time step the highest activated node in the model was selected as a candidate answer to the remote associate test. If the answer matched the target then the model's activation cycling was stopped. If the candidate answer did not match the target then the activation of this node was temporarily set to zero, and activation cycled in the network for a further time step. The number of time steps taken to activate the answer to the highest level within the network was taken as a measure of the problem's difficulty. Maximum activation of nodes in the network was set at 1.

Preliminary analyes of the model's performance have shown that it is able to simulate distinctions between easy and hard remote associates tests in terms of time taken to solve the problem in the model, as well as reflecting behavioural solution rates from studies demonstrating a range of problem difficulty[19], and studies manipulating easy versus difficult problems.[20] Focusing in on this latter study, in behavioural studies problems were distinct in terms of their solution success rate, and the difficulty was predicted to a certain extent by the nearness of the association between the stimulus words and the target word. However, the model was better able to predict solution difficulty as a property of the network, rather than single connections between units. Harder problems were those where the target was in a small world other than those in which the stimulus words occurred. Figure 1 shows an example of a difficult (upper) and easy (lower) problem. For the easy problem, the target occurs with connections to and from each of the small worlds that contain the stimulus words. In addition, the stimulus words' small worlds also inter-associate, meaning that activation can quickly build up within the region of the semantic space that is occupied by the target word.

In contrast, for the difficult problem, the target word occurs in a small world relatively distinct from the small worlds containing the stimulus words. Thus, activating the target word to be more active than closely associated words to the stimulus words takes considerably longer than for the easy problem.

The model demonstrates not only quantitative effects of remote associates problem solving, but also qualitative features. For instance, Moss[11] transcribed participants' attempts to find the answer to a remote associates test, and reports that participants often become fixated on one or two potential answers that they find difficult to escape. Our model demonstrates similar effects. The highest activated word in the network is selected as a potential target, but then its activation is temporarily reduced, and the next highest activated word is selected. But then, for words that are highly-associated and highly-interconnected to other activated words the temporary reduction of activation is replenished quickly within the model (see Dodds, Smith and Ward[5] for an experimental manipulation of this property). For difficult problems, such as the lower diagram in Figure 1, the model can spend several steps jumping between words appearing in one small world containing a stimulus word, and then another small world containing one of the other stimulus words. Only when activation has spread sufficiently widely within the model will the target be available for selection as a possible solution.

5. Conclusion

We have described some of the impediments to computational models of problem solving, particularly focusing on the role of the activation of semantic information in memory. We have made the case that by attending to the small world structure of semantic memory relevant to problem solving, an implemented computational model can bypass many of these apparent obstacles. In particular, spreading activation can be observed over long time periods, and traces of particular problems in incubation effects can be maintained without catastrophic interference between separate problems. Hence, because of the small world structure, activation of unsolved problems – as in the Zeigarnik effect – can be simultaneously maintained without activation within the model spiraling out of control, or confusion resulting from too much overlap between semantic networks of activations related to individual problems. Observing the structure of associative memory can thus simplify the modelling architecture required in order to account for a broad range of problem solving performance. In return, merging semantic memory with problem solving accounts opens up a whole body of experimental effects in problem solving which can provide insight into the function and operation of semantic memory.

Finally, by now, we hope the reader has had enough time to incubate the problems we presented in section 1.2 in order to access the small worlds inhabited by the answers. However, if the answers are still escaping the reader, we provide them now. Easy problem: *broken, clear, eye*; answer: *glass*. Difficult problem: *high, book, sour*; answer: *note*.

References

1. E. M. Bowden and M Jung-Beeman, *Behavioral Research, Methods, Instruments, and Computers* **35**, 634 (2003).
2. D. J. Cai, S. A. Mednick, E. M. Harrison, J. C. Kanady and S. C. Mednick, *Proceedings of the National Academy of Sciences, USA* **106**, 10130 (2009).
3. C. Chiarello, C. Burgess, L. Richards and A. Pollock, *Brain and Language* **38**, 75 (1990).
4. A. M. Collins and E. F. Loftus, *Psychological Review* **82**, 407 (1975).

5. R. A. Dodds, S. M. Smith and T. B. Ward, *Creativity Research Journal* **14**, 287 (2002).
6. J. B. T. Evans, *Annual Review of Psychology* **59**, 255 (2008).
7. S. Helie and R. Sun, Psychological Review **117**, 994 (2010).
8. T. T. Hills, M. N. Jones and P. M. Todd, Psychological Review **119**, 431 (2012).
9. J. L. McClelland, *Proceedings of the Third Annual Meeting of the Cognitive Science Society*, 170 (1981).
10. S. A. Mednick, *Psychological Review* **69**, 220 (1962).
11. J. Moss, Unpublished doctoral dissertation, Carnegie Mellon University (2006).
12. D. L. Nelson, C. L. McEvoy and T. A. Schreiber, *The University of South Florida word association, rhyme, and word fragment norms* (1998).
13. S. Ohlsson, In M. T. Keane & K. J. Gilhooly (Eds.), *Advances in the psychology of thinking*, 1. London: Harvester Wheatsheaf (1992)
14. S. Ohlsson, AAAI Spring Symposium: Creative Intelligent Systems, 61 (2008).
15. M. R. Quillian, *Mechanical Translation* 7, 17 (1962).
16. J. G. W. Raaijmakers and R. M. Shiffrin, *Psychological Review* **88**, 93 (1981).
17. S. J. Russell and P. Norvig, *Artificial intelligence: A modern approach (2nd ed.)*. Upper Saddle River, New Jersey: Prentice Hall (2003).
18. M. Schilling, *Creativity Research Journal* **17**, 131 (2005).
19. Smith, S. M. and Blankenship, S. E., *American Journal of Psychology* **104**, 61 (1991).
20. U. N. Sio, P. Monaghan and T. Ormerod, *Memory and Cognition* **41**, 159 (2013).
21. U. N. Sio and T. C. Ormerod, *Psychological Bulletin* **135**, 94 (2009).
22. S. M. Smith, In S. M. Smith, T. B. Ward, & R. A. Finke (Eds.). *The Creative Cognition Approach*, 135. Cambridge, Mass.: MIT Press. (1995)
23. M. Steyvers and J. B. Tenenbaum, *Cognitive Science* **29**, 41 (2005).
24. B. V. Zeigarnik, In W. D. Ellis (Ed.), *A sourcebook of Gestalt psychology*. New York: Humanities Press. (1927, 1967).

ON OBSERVATIONAL LEARNING OF HIERARCHIES IN SEQUENTIAL TASKS:
A DYNAMIC NEURAL FIELD MODEL

EMANUEL SOUSA

Department of Industrial Electronics, University of Minho, Guimarães, Portugal
E-mail: esousa@dei.uminho.pt

WOLFRAM ERLHAGEN

Centre of Mathematics, University of Minho, Guimarães, Portugal;
Donders Institute for Cognition Brain and Behavior,
Radboud University Nijmegen, The Netherlands
E-mail: wolfram.erlhagen@math.uminho.pt

ESTELA BICHO

Department of Industrial Electronics, University of Minho, Guimarães, Portugal
E-mail: estela.bicho@dei.uminho.pt

Many of the tasks we perform during our everyday lives are achieved through sequential execution of a set of goal-directed actions. Quite often these actions are organized hierarchically, corresponding to a nested set of goals and subgoals. Several computational models address the hierarchical execution of goal-directed actions by humans. However, the neural learning mechanisms supporting the temporal clustering of goal-directed actions in a hierarchical structure remain to a large extent unexplained. In this paper we investigate in simulations, of a dynamic neural field (DNF) model, biologically-based learning and adaptation mechanisms that can provide insight into the development of hierarchically organized internal representations of naturalistic tasks. In line with recent experimental evidence from observational learning studies, the DNF model implements the idea that prediction errors play a crucial role for grouping fine-grained events into larger units. Our ultimate goal is to use the model to endow the humanoid robot ARoS with the capability to learn hierarchies in sequential tasks, and to use that knowledge to enable efficient collaborative joint actions with human partners. For testing the ability of the system to deal with the real-time constraints of a learning-by-demonstration paradigm we use the same assembly task from our previous work on human-robot collaboration. The model provides some insights on how hierarchically structured task representations can be learned and on how prediction errors made by the robot and signaled by the demonstrator can be used to control this process.

1. Introduction

In our everyday activities such as cleaning the kitchen, preparing a meal or serving a coffee a complex stream of object-directed actions has to be organized in real time in order to fulfill the task. Although these actions unfold in time as a linear sequence, goal-directed behavior is very often organized hierarchically, corresponding to a nested set of goals and subgoals.[1-3] Groups of goal-related actions appear to be linked together at multiple timescales through their relations to a common cause and not primarily through their temporal and causal relations to each other. Such a hierarchical organization of behavior is believed to support efficient action planning since it allows the actor to select individual actions and action clusters based on their predicted consequences in the near and distal future. It also promotes flexibility in task execution since the temporal order of subgoals and their associated subroutines may be easily adapted to changing environmental constraints. Think of the coffee serving example. The ingredients coffee, milk and sugar can in principle be added in an any order, but all subgoals have to be completed before the cup can be handed over.

Several existing computational accounts explain important aspects of hierarchical action production in humans (e.g. Refs. 4 and 5). However, the neural learning mechanisms that give rise to the temporal clustering of goal-directed actions in a hierarchical structure remains to a large extent unexplained. Developmental studies with children reveal that infants at a surprisingly early age of 24-36 months are able to imitate the goal-subgoal structure of novel action sequences composed of several separate, but jointly necessary means steps.[6,7] In other words, the toddlers are not simply copying the chain of events demonstrated by the adult but imitate on a hierarchical level. In these studies, the children had little explicit event knowledge that might help them to structure the task from the beginning on, suggesting the operation of bottom-up mechanisms in learning the hierarchical structure.

In the present paper, we investigate in simulations of a dynamic neural field (DNF) model biologically-based learning and adaptation mechanisms that can provide insight into the development of hierarchically organized internal representations of naturalistic tasks. Specifically, we explore the idea that prediction errors may play a fundamental role in grouping fine-grained events into larger units.[8] Ultimately, we are interested in using the DNF model to endow the humanoid robot ARoS developed by our group with a basic hierarchy imitation competence. It is thus important that the model is able to capture the real-time constraints of a learning by demonstration paradigm.[9] To this end, we adopt for the model simulations an assembly task from our previous DNF approach to natural human-robot interactions[10,11] for a socially guided learning paradigm.[12] One or more human teachers first demonstrate possible linear sequences of assembly steps that result in the construction of a toy vehicle from its parts. During observation, the dynamic field model establishes synaptic links between sufficiently active neural populations encoding the perceived consequences of subsequent assembly steps using Hebbian learning principles. The robot then tries to reproduce the sequential task and makes at every stage of the assembly work a prediction about the subgoal that has to be achieved next. In the socially guided learning paradigm, the teacher gives immediate positive or negative feedback about this prediction. If a prediction failure occurs, the error signal is used to lower the threshold for associative learning during new task demonstrations.[13] As a result, time-dependent population activity representing the memory of already accomplished subgoals remains above threshold for a longer time period. The system is thus able to develop longer-term sequential dependencies that reduce the likelihood of prediction failure. In addition, the error signal becomes associated with the segmentation of the assembly work in independent subtasks.[8,14,15] A neural population, which is initially driven by the error signal, establishes connections to the cluster of all subgoal representations defining a certain subtask.

Learning of task hierarchies thus necessarily means that the activity of the higher-level representation has to be maintained above the learning threshold during the course of the sequential activation of the subgoal representations. Dynamic neural field theory provides a rigorous mathematical framework to explain the existence of self-stabilized activity patterns of neural populations representing such a working memory function.[16,17]

2. Dynamic Neural Field Model

The model is based on the theoretical framework of Dynamic Neural Fields (DNFs) that was originally proposed as a simplified mathematical model to explain firing patterns of neuronal populations in cortical tissue.[16,18] The architecture of this model family reflects the hypothesis that strong recurrent excitatory and inhibitory interactions in local populations of neurons form a basic mechanism for cortical information processing. The recurrent interactions cause non-trivial dynamic behaviour in neural assemblies. Most importantly, population activity which is initiated by time-dependent external signals may become self-sustained in the absence of any external

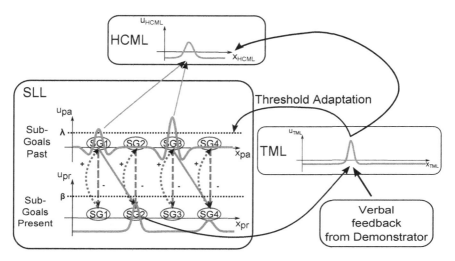

Fig. 1. Schematic of the model architecture with three connected layers implementing sequence learning (SLL), task monitoring (TML), and higher level cognitive memory (HCML).

input. Such attractor states of the population dynamics may be used by the nervous system to guide complex goal-directed behavior that goes beyond simple input-output mappings. DNF models have been used in the past to model cognitive capacities like working memory, decision making, prediction and anticipation[19,20] and to synthesize these functionalities in artificial agents (for a review see Ref. 17).

Figure 1 presents a sketch of the model architecture for learning and representing hierarchical structure in sequential tasks. It consists of three layers with neural populations that receive input from connected neural pools and external sources (vision, speech). In the sequence learning layer (SLL), a Hebbian mechanism is applied to store the temporal order(s) of subgoals as demonstrated by the teacher. The task monitoring layer (TML) contains populations that become active when during task execution an erroneous prediction about a subsequent assembly step is detected. Suprathreshold population activity in TML triggers in turn the development of population activity in the higher cognitive memory layer (HCML) that signals the accomplishment of an entire subtask composed of several subgoals represented in SLL.

The model takes inspiration form recent findings in cognitive neuroscience. Neural populations in areas of the prefrontal cortex (PFC) are known to be concerned with cognitive aspects of behavioral planning.[21]

They seem to encode the end result of object-directed behavior (e.g., a change in the state of the target object) rather than the intended movements required to perform the task. We conceptualize this finding in our model by assuming that distinct neural populations encode the desired end-states or goals of individual assembly steps. For successful execution of a sequential task, the working memory of already completed steps has to be separated from steps that are still to be accomplished. Firing patterns of neural populations in PFC suggest that previous and future goals are indeed represented by separate but interacting neuronal assemblies.[22] The DNF-model reflects this finding by assuming the existence of a "past" layer, u_{pa}, and a "present" layer, u_{pr}, each with representations of end-states of all assembly steps. Populations encoding the same subgoal are coupled by pre-defined excitatory and inhibitory connections (see Fig. 1, dotted lines in SLL). During task demonstration, vision input indicating a change in the state of the target object initiates an interaction dynamics that results in an automatic updating of the working memory in u_{pa}. Subsequently, a Hebbian learning mechanism establishes connections to populations in u_{pr} representing newly demonstrated subgoals (see Fig. 1, solid arrows in SLL) as long as the population activity in u_{pa} remains above the learning threshold[23] (for a discussion of the threshold concept see Ref. 13). Figure 2 compares the time course of population activity in u_{pr} and u_{pa} when at time t_0 the population in u_{pr} receives input from the vision system. The input triggers a supra-threshold activity pattern that in turn starts to drive through the pre-defined excitatory connections the corresponding population in u_{pa}. The activity in u_{pa} continuous to grow due to the excitation from u_{pr}, the vision input and the recurrent interactions within the population. The inhibitory feedback connections cause in turn a decrease of the activity in u_{pr} to resting level, resulting in a transient activity pattern in layer u_{pr}. Also the activity in u_{pa} decreases to some extent over time. However, the recurrent interactions within the populations are strong enough to self-stabilize the activity well above resting level, thus establishing a memory function. The learning threshold λ defines the time window $[t_3, t_5]$ of high activation in which new associations can be establishes to sufficiently active populations ($u_{pr} > \beta$) representing newly demonstrated subgoals.

For executing the sequential task, the robot has to switch from an observational learning mode to an active mode in which at each stage of the construction the robot generates itself a prediction about possible subgoals. To realize this switch, we exploit the finding in neurophysiological and computational studies showing that a task-dependent change in baseline acti-

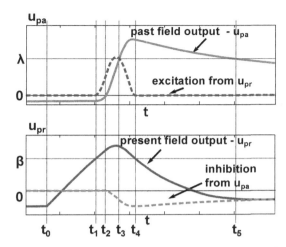

Fig. 2. Time course of activity (solid lines) and input (dashed lines) of two interconnected populations in u_{pr} and u_{pa} encoding the same subgoal.

vity may have a profound influence on the integrative properties of neuronal populations.[24,25] A higher resting level in the active mode makes it possible that the visual input alone drives the populations in u_{pa} to a persistent state. Through the newly established connections, populations in layer u_{pr} become subsequently activated above resting level. Competition between different subgoal representations mediated by lateral inhibition ensures that only one population reaches a suprathreshold activation level at the same time. In the real-word robotics implementations, the robot verbally communicates this prediction about the next subgoal and the human teacher gives immediate positive or negative feedback. Negative feedback signalling a prediction error (e.g., a premature execution of a certain subgoal) shapes the learning process in two ways. First, the verbal input together with input from the active population in u_{pr} creates a self-stabilized activity pattern of a specific population in the task monitoring layer (TML). We generalize here the notion of a comparison between population representations relevant for correct task execution implemented in our previous dynamic field model of natural human-robot interactions.[11] A "generic" error detection system seems to be sensible not only to internal but also to external sources such as feedback.[26] Suprathreshold activity in TML causes an adaptation of the gating threshold λ for Hebbian learning during new demonstrations of the task.[13] Functionally, a lower threshold increases the time window for learning in u_{pa}, allowing the formation of associations to several subsequent subgoal representations in u_{pr}.

A second important role of the prediction failure at the beginning of the learning is that it introduces a breakpoint in the linear processing of the sequential task. It has been hypothesized that this breakpoint might be exploited by the learning system to "chunk" groups of subgoals together to form a higher-level subtask representation.[8,14,27] Such a strategy might provide the basis of hierarchical processing. We take inspiration from models of cognitive control that attribute different levels of temporal abstraction in sequential actions to different areas in the frontal lobe.[3,28] In the model, the population activity in TML enables the spontaneous creation and maintenance of a suprathreshold activation pattern of a population in the higher cognitive memory layer (HCML) by providing homogeneous excitatory input to this population. The pattern becomes meaningful during the learning process since the Hebbian principle establishes connections to all subgoal representations defining the subtask. The process of creating the higher level cognitive memory stops when the sustaining homogeneous input from TML disappears. This happens to occur when positive verbal feedback during task execution destabilizes the population representation of the initial prediction error.

2.1. *Mathematical details*

Each layer of the model is formalized by a DNF. The temporal evolution of activity, $u(x_i, t)$, of a neuron at position x in field i is governed by a particular form of a dynamic neural field:

$$\tau_i \frac{\partial u(x_i, t)}{\partial t} = -u(x_i, t) + \int w_i(x_i - x') f_0 [u(x', t)] \, dx' + h_i + S(x_i, t) \quad (1)$$

where $\tau_i > 0$ is the time constant of the dynamics and $h_i < 0$ defines the baseline level to which field activity relaxes without external input. The integral term describes the intra-field interactions to which only sufficiently active neurons contribute. The non-linear transfer function f_0 is chosen as the Heaviside step function with threshold 0.

The input $S(x_{pa}, t)$ to field $u(x_{pa}, t)$ is given by:

$$S(x_{pa}, t) = C_{pa} I(x, t) + \int f_\beta [u(x_{pr}, t)] G(x_{pr}, x_{pa}) \, dx_{pr} \quad (2)$$

where $I(x, t)$ represents Gaussian input provided by vision system with a strength parameter C_{pa}. The integral term represents the summed input from $u(x_{pr}, t)$ mediated by pre-defined excitatory connections $G(x_{pr}, x_{pa})$ between corresponding populations. The threshold function $f_\beta [u(x_{pr}, t)]$

ensures that only activity above β is propagated. For the field $u(x_{pr}, t)$ the input $S(x_{pr}, t)$ is given by:

$$S(x_{pr}, t) = C_{pr}I(x, t) + \int f_0\left[u(x_{pa}, t)\right] J\left(x_{pa}, x_{pr}\right) dx_{pa} + \\ + \int f_0\left[u(x_{pa}, t)\right] a\left(x_{pa}, x_{pr}, t\right) dx_{pa} \tag{3}$$

where $S(x_{pr}, t)$ consists of three components: the input from the vision system $I(x, t)$, the input from $u(x_{pa}, t)$ mediated by the pre-defined inhibitory connections $J(x_{pa}, x_{pr})$ and the summed activity from populations in $u(x_{pa}, t)$ mediated by the adaptive excitatory connections $a(x_{pa}, x_{pr}, t)$ established during the learning process. The mathematical formulation of the learning rule for setting the synaptic connections $a(x_{pa}, x_{pr}, t)$, between any two sufficiently active neurons $u(x_{pa}, t) > \lambda$ and $u(x_{pr}, t) > \beta$ is defined by:

$$\tau_a \frac{\partial a(x_{pa}, x_{pr}, t)}{\partial t} = f_\beta[u(x_{pr}, t)]f_\lambda\left[u(x_{pa}, t)\right] e(x_{pr}, t) \tag{4}$$

It represents a supervised Hebbian Leaning rule (see Ref. 29) where the parameter τ_a defines the time scale of learning and the error term $e(x_{pr}, t)$ describes the difference between the desired active state of a neuron in $u(x_{pr}, t)$ and the summed input from all connected neurons in $u(x_{pa}, t)$:

$$e(x_{pr}, t) = f_\beta[u(x_{pr}, t)] - \int f_\lambda\left[u(x_{pa}, t)\right] a(x_{pa}, x_{pr}, t) dx_{pa} \tag{5}$$

3. Modeling Results

The main objective of the present modeling work is to show that the neuro-plausible learning mechanisms in a social teaching context are in principle sufficient to endow the robot ARoS with the capacity to learn and represent generalized task knowledge in a hierarchical organization. For our learning by demonstration approach it is important to stress that ARoS already has the necessary perceptual and motor capacities (for details see Refs. 10 and 11): the different reach-grasp-place sequences to achieve each individual assembly step are in its motor repertoire, the vision system is trained to recognize the outcome of each assembly step, and a speech syntheziser/recognizer system endows the robot with elementary verbal communication skills.

ARoS has to learn the construction of a toy vehicle consisting of a bottom platform (BP) with an axle on which two wheels (LW, RW) have first to be mounted and then fixed with nuts (LN, RN). Four columns (C1, C2, C3, C4) have to be inserted into specific holes on the bottom platform before a top platform (TP) can be attached (Fig. 3). The human teachers are instructed to show in each demonstration the whole sequential task, but are allowed to vary the sequential order. As a concrete example, we focus in the present modeling work on learning a specific subtask: C1 to C4 may be inserted in any order, but TP requires the presence of all columns.

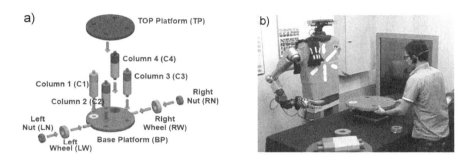

Fig. 3.   (a) Toy vehicle (b) Robot ARoS interacting with a human teacher.

3.1. Demonstrating of two sequential orders

In the first experiment, two teachers demonstrate the assembly task using two sequences of subgoals with the ordinal position of C3 and C4 interchanged. Each of the two sequences were demonstrated 15 times in alternating order. The recall trials start with the presentation of BP by the teacher and the vision input activates the respective population in u_{pa}. In line with the training sequences, the model predicts C1 as the next subgoal. After having inserted C2, the robot will predict C3 and C4 as equally likely future assembly steps. Small differences in the learned weights and the population activity due to noise in the system gives preference to one of the options. In the example of Fig. 4, the population representing C4 becomes suprathreshold. In any case, the robot always predicts the last missing column as new subgoal before trying to place the top platform. This successful imitation of the demonstrated sequences does not mean, however, that the robot has already understood the causal relationship between the columns and the top platform. The following experiment shows this.

3.2. *Demonstration of a new sequential order*

Now the new subsequence $BP \rightarrow C4 \rightarrow C3 \rightarrow C1 \rightarrow C2 \rightarrow TP$ is demonstrated in alternation with the two previously demonstrated sequential orders (15 presentations for each order). During task execution, the robot follows the new sequential order. However, after having correctly predicted

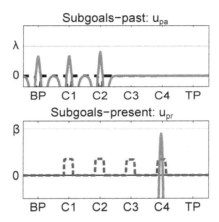

Fig. 4.   Population activity (solid lines) in the past layer u_{pa} and in the present layer u_{pr}. The dashed line in u_{pr} represents the input from active populations in u_{pa} through learned connections $a(x_{pa}, x_{pr}, t)$. The robot makes the prediction to insert C4 after having finished subgoals BP, C1 and C2.

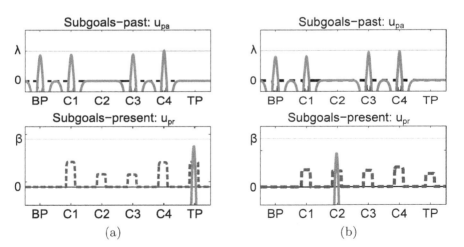

Fig. 5.   (a)Wrong prediction of TP given that C2 is still missing. (b) Correct prediction of C2 following the learning of longer-term dependencies during new demonstrations.

and inserted C4, C3 and C1, the robot makes the prediction to attach TP instead of inserting C2 (Fig. 5(a)). This premature choice is a direct consequence of the simple associative chaining that the model has created during task demonstration. The TP population gets direct input from the population C4 and C3 (sequences 1 and 2) whereas C2 is supported only by population C1 (sequence 3). In response to the negative verbal feedback, a self-sustained activation peak in TML evolves (not shown) representing the prediction error in the self-generated sequence. This activation is assumed to lower the adaptive learning threshold λ in u_{pa} by a fixed amount,[13] thus increasing the time window for associative learning. As a consequence, connections to several subsequent subgoal representations may be learned during new demonstration trials. The threshold adaptation in response to negative feedback may continue for several demonstration-execution iterations until positive feedback indicates a successful execution of the whole subsequence. This is shown in Fig. 5(b) where now C2 follows C1, C3 and C4. Note that also TP continues to get input from active populations in u_{pa}, which is however weaker compared to the input to C2. The suprathreshold activity in TML also triggers and sustains the activation of a pool of neurons in HCML (compare the time course in Fig. 6(a) on top). During execution trials, subgoal representations in the past layer u_{pa} become associated with this highly active population. When the prediction error is corrected, that is, the activity in TML has decayed to resting level, all populations defining the subtask of finishing the base with 4 columns (TB plus C1 to C4) are active and thus drive the population in HCML. This is shown in Fig. 6(b) when at the time of placing C2, suprathreshold activity in HCML evolves. Note the delay of temporal evolution compared to the time course of suprathreshold activity of population C2. This model prediction is in line with experimental findings in sequence learning studies showing that it takes some extra time to mentally mark the boundaries of the higher level subtask when the associated last subgoal has been achieved.[27]

4. Discussion

Sequentially organized human activity, such as assembling an object, can be identified at a range of different time-scales, from fine-grained to coarse-grained. The segmentation on different temporal grains reflects a hierarchical organization of behavior where goals tend to be satisfied by the recursive satisfaction of subgoals. Recent experimental evidence from observational learning studies suggests that being able to encode the hierarchical structure of observed activity strongly promotes the successful imitation of new

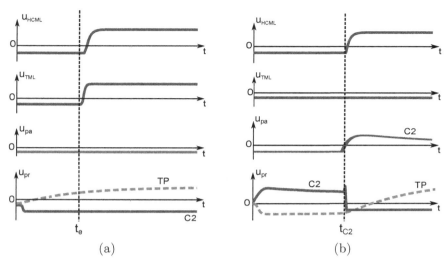

Fig. 6. Time course of activity in the various model layers (a) Error trial: Population TP in u_{pr} becomes active before the last column (C2) has been inserted. At time t_e, the human gives negative feedback to this prediction, triggering the suprathreshold activity first in TML and subsequently in HCML. (b) Correct trial: After learning, the system correctly recalls C2 before TP. C2 is inserted at time t_{C2}, TP is predicted and the HCML population indicating the completion of the subtask becomes activated.

action sequences (e.g. Refs. 7 and 27). Learning by observation and imitation has also attracted a lot of attention over the last couple of years in the robotics community since it represents an intuitive and efficient way how ordinary people could teach a robot new skills and tasks.[9,30] In this paper we have addressed the crucial question, in a learning by observation paradigm, of how an agent with no prior task knowledge (e.g., a robot or a child) may learn to segment the continuous flow of sequentially structured behavior into a goal-subgoal hierarchy. In line with recent experimental evidence, the dynamic field model implements the idea that prediction errors play a crucial role for grouping fine-grained events (e.g., individual assembly steps) into larger units (e.g., the subtask of inserting 4 columns). An adaptation of the learning threshold triggered by the prediction error updates the long-term sequential dependencies (stored in the learned inter-field connections) necessary to achieve accurate prediction. The mechanism is similar to the gating signal that Reynolds et al.[15] have used in their connectionist network model of event segmentation to learn and update internal context information reflecting event knowledge. They applied their segmentation model to perform one-step predictions of human movement patterns. This suggests that prediction errors might be sufficient to learn part-subpart structures simultaneously on different levels of the action hierarchy. In the social learning situation of the robotics experiments, the human teacher gives immediate binary feedback when the robot predicts the next subgoal of the assembly task. This greatly improves the efficiency of the learning compared to reinforcement-based mechanisms since the credit-assignment problem in complex sequences is avoided. Interestingly, integrating the notion of temporal abstract actions associated with (pseudo-)rewarded subgoals into the reinforcement learning framework can significantly speed up learning (for discussion see Ref. 3).

A unique feature of the model is that the time window for Hebbian learning is defined by the time course of the population activity above learning threshold. This means that only the initial stage of working memory (WM) maintenance of already accomplished subtasks contributes to successful long-term memory (LTM) formation of sequential order. This model prediction is in line with recent findings in combined behavioral and brain imaging studies of brain circuits commonly believed to support the transition from WM and LTM.[31]

The capacity to form hierarchies of sequential behavior is not only important for successful imitation. Compact representations of temporally extended activity are also beneficial to plan future actions and to coordinate with others in joint action tasks.[11] In the present implementation, the population activity in HCML encodes that a certain subtask has been completed. We are currently working on an extended version of the learning model which basically copies the two layered structure of the subgoal level to allow a sequential activation of future subtask representations.

Acknowledgements

Work supported by: (1) Portuguese FCT Grant SFRH/BD/48529/2008, financed by POPH-QREN-Type 4.1-Advanced Training, co-funded by the European Social Fund and national funds from MEC; (2) FEDER Funds through Competitivity Factors Operational Program - COMPETE and National Funds by FCT Portuguese Science and Technology Foundation under the Project FCOMP-01-0124-FEDER-022674. (2) Project NETT: Neural Engineering Transformative Technologies, EU-FP7 ITN proj. nr. 289146

References

1. K. Lashley, The problem of serial order in behavior, in *Cerebral Mechanisms in Behaviour*, ed. L. A. Jeffres (Wiley, New York, 1951) pp. 112–142.

2. C. A. Kurby and J. M. Zacks, *Trends in Cognitive Sciences* **12**, 72 (2008).
3. M. M. Botvinick, *Trends in Cognitive Sciences* **12**, 201 (2008).
4. R. P. Cooper and T. Shallice, *Psychological Review* **113**, 887 (2006).
5. M. M. Botvinick and D. C. Plaut, *Psychological Review* **111**, 395 (2004).
6. L. L. Travis, Goal-based organization of event memory in toddlers, in *Developmental spans in event comprehension and representation: Bridging fictional and actual events*, (Lawrence Erlbaum Associates, Inc, 1997) pp. 111–138.
7. A. Whiten, E. Flynn, K. Brown and T. Lee, *Developmental Science* **9**, 574 (2006).
8. J. M. Zacks, C. A. Kurby, M. L. Eisenberg and N. Haroutunian, *Journal of Cognitive Neuroscience* **23**, 4057 (2011).
9. W. Erlhagen, A. Mukovskiy, E. Bicho, G. Panin, C. Kiss, A. Knoll, H. T. van Schie and H. Bekkering, *Robotics and Autonomous Systems* **54**, 353 (2006).
10. E. Bicho, W. Erlhagen, L. Louro and E. C. e Silva, *Human Movement Science* **30**, 846 (2011).
11. E. Bicho, W. Erlhagen, L. Louro, E. C. e. Silva, R. M. Silva and N. Hipólito, A dynamic field approach to goal inference, error detection and anticipatory action selection in human-robot collaboration, in *New Frontiers in Human-Robot Interaction*, eds. K. Dautenhahn and J. Sanders (John Benjamins Publishing Company, 2011) pp. 135–164.
12. A. Thomaz and C. Breazeal, *Artificial Intelligence* **172**, 716 (2008).
13. A. R. Seitz and H. R. Dinse, *Current Opinion in Neurobiology* **17**, 148 (2007).
14. K. Sakai, K. Kitaguchi and O. Hikosaka, *Experimental Brain Research.* **152**, 229 (2003).
15. J. R. Reynolds, J. M. Zacks and T. S. Braver, *Cognitive Science* **31**, 613 (2007).
16. S.-i. Amari, *Biological Cybernetics* **27**, 77 (1977).
17. W. Erlhagen and E. Bicho, *Journal of Neural Engineering* **3**, R36 (2006).
18. H. R. Wilson and J. D. Cowan, *Kybernetik* **13**, 55 (1973).
19. G. Schöner, *Cambridge Handbook of Computational Cognitive Modeling* , 101 (2007).
20. W. Erlhagen and G. Schöner, *Psychological Review* **109**, 545 (2002).
21. J. Tanji, K. Shima and H. Mushiake, *Trends in cognitive sciences* **11**, 528 (2007).
22. A. Genovesio, P. J. Brasted and S. P. Wise, *The Journal of Neuroscience: The Official Journal of the Society for Neuroscience* **26**, 7305 (2006).
23. D. O. Hebb, *The organization of Behavior* (Lawrence Erlbaum Associates Inc, Cambridge, MA, USA, 1949).
24. E. Salinas, *Neural Computation* **15**, 1439 (2003).
25. W. Asaad, G. Rainer and E. Miller, *Journal of Neurophysiology* **84**, 451 (2000).
26. W. H. R. Miltner, C. H. Braun and M. G. H. Coles, Event-related brain potentials following incorrect feedback in a time-estimation task: Evidence for a generic neural system for error detection (1997).

27. B. M. Hard, S. C. Lozano and B. Tversky, *Journal of Experimental Psychology. General* **135**, 588 (2006).
28. J. Fuster, *International Journal of Psychophysiology* **60**, 125 (2006).
29. P. Dayan and L. Abbott, Pasticity and learning, in *Theoretical Neuroscience: Computational and Mathematical Modeling of Neural Systems*, (MIT Press, Cambridge, MA, USA, 2001)
30. C. L. Dautenhahn, Kerstin and Nehaniv (ed.), *Imitation in Animals and Artifacts* (MIT Press, Cambridge, MA, USA, 2002).
31. C. Ranganath, M. X. Cohen and C. J. Brozinsky, *Journal of Cognitive Neuroscience* **17**, 994 (2005).

KNOWING WHEN TO QUIT ON UNLEARNABLE PROBLEMS: ANOTHER STEP TOWARDS AUTONOMOUS LEARNING

T. R. SHULTZ and E. DOTY

McGill University
Montreal, Canada

Autonomous learning is the ability to learn effectively without much external assistance. An important strength of autonomous learners is that they can shape their own learning and development, in large part by choosing which problems to learn. Such choices include selecting a problem to learn and deciding whether to continue learning on that selected task or abandon it in favor of something else. We extend a constructive neural-learning algorithm, sibling-descendant cascade-correlation, to monitor lack of progress in learning so that unproductive learning can be abandoned. Learning is abandoned when network error fails to change by more than a specified threshold for a specified number of consecutive learning cycles. Here we explore the space defined by these threshold and patience parameters on problems of different degrees of learnability. Our results simulate findings from recent experiments with infants who abandon learning on difficult tasks and focus their attention on tasks of moderate difficulty.

1. Introduction

Autonomy is a desired quality in machine learning and artificial intelligence where effectiveness is compromised whenever human intervention is required. It is also a characteristic of biological agents as they learn in their natural environments. Autonomous learners can, to some extent, shape their own learning and development, in large part by choosing what problems to work on. Such choices include selecting a problem to learn and deciding whether to continue learning on the selected task or abandon it in favor of something else.

Knowing when to stop learning has two aspects – quitting either when a problem has been mastered and when it is not likely to be mastered. In the constructive neural networks that we use, typically victory is declared and learning is terminated when the network is correct on all training examples, meaning that its outputs are within some score-threshold of their target values. [1,2]

Quitting without mastery is considerably more problematic, but can be analyzed in terms of costs and benefits. The total cost of learning can be conceptualized as energy expenditure (of the learning effort) plus opportunity cost (the value of the best alternative not chosen, whether other learning or exploitation of existing knowledge): $Cost_{Total} = Energy_{Learn} + Cost_{Opportunity}$.[3] The net payoff of learning is the benefit of successful learning minus the total cost of learning: $Payoff_{Net} = Benefit_{Learn} - Cost_{Opportunity}$ When an agent continues working on an unlearnable problem, there is a large negative payoff: considerable cost without any benefit. Having started learning a difficult problem, it can be sensible to abandon it as soon as lack of progress becomes evident.

Past computational modeling suggests that a key factor in early learning cessation is whether progress is being made.[4] In that work, learning progress was monitored by tracking the first derivative of error reduction to identify intrinsic rewards, while a reinforcement-learning module selected actions to maximize these rewards. There was also an external network to assess learning progress.

Similarly, in other computational work, error in the near-past was subtracted from error in the far-past to assess learning progress.[5,6] In that research, there were also modules for probabilistic action selection and progress assessment at different levels.

In psychological research, infant looking time was modeled by the information-theoretic properties of stimuli.[7] The negative log probability of an event, corresponding to the bits of information in a stimulus, was conditioned on previous events. The larger the negative log probability, the more surprising the current event should be. As that model predicted, infants were more likely to look away from either highly informative or uninformative events. The authors referred to this as the *Goldilocks* effect because infants preferred to look at stimuli that were neither too easy nor too difficult, but had moderate degrees of complexity.

Other experiments reported that infants attend longer to learnable as opposed to unlearnable artificial-language grammars, taking more trials and more time on grammars in which a valid generalization over examples could be made.[8] Thus, there is independent evidence that infants use an implicit metric of their learning progress to direct their attention to learnable material.

We extended a constructive neural-learning algorithm, sibling-descendant cascade-correlation (SDCC)[9], to monitor lack of progress in learning in order to autonomously abandon unproductive learning.[3] The extended algorithm abandoned a learning task when network error failed to change by more than a specified amount (threshold) for a specified number of consecutive learning cycles (patience). These networks simulated results of the foregoing infant experiments, abandoning the task when learning succeeded and when lack of progress became evident.

This extended algorithm also avoided network overtraining effects in a more realistic manner than conventional use of validation test sets.[3] Such validation test sets are not apparent in natural learning, where the only error signals occur during learning. Rather, validation test sets are programmer aids to detect that training has gone on too long, signaled by rising error on test problems even as error continues to drop on the training patterns.[10,11] The idea is that networks with rising test error are starting to memorize training patterns instead of making abstractions that will be useful in generalizing to untrained examples. Our extended algorithm avoided such overlearning, not with validation test sets, but by raising the learning-threshold value. In this paper, we more systematically explore the space defined by the patience and threshold parameters in larger-scale experiments.

2. Method

2.1. *Cascade-correlation*

Among the prominent constructive neural-network algorithms are Cascade-correlation (CC) and SDCC. CC networks start with just input and output units, typically fully connected. They are feed-forward networks trained in a supervised fashion with examples of particular input and target-output values. Hidden units required to deal with non-linearity in the training patterns are recruited one at time, as needed. The algorithm alternates between output and input phases to reduce error and recruit helpful hidden units, respectively.[1] The function to minimize error during output phase is:

$$E = \sum_{o} \sum_{p} (A_{op} - T_{op})^2 \qquad (1)$$

where E is error at the output units, A is actual output activation, and T is target output activation for unit o and pattern p. Error minimization is accomplished with the *Quickprop* algorithm, a fast variant of the generalized delta rule that uses curvature as well as slope of the error surface to compute weight changes.[12] When error can no longer be reduced by adjusting weights entering the output units, the CC algorithm switches to input phase to recruit a hidden unit to supply additional computational power.

In the input phase, a pool of eight candidate hidden units is established, typically with sigmoid activation functions, and with initially random but trainable weights from the input units and any existing hidden units. These weights are trained by attempting to maximize a covariance C between candidate-hidden-unit activation and network error:

$$C = \frac{\sum_o | \sum_p (h_p - \langle h \rangle)(e_{op} - \langle e_o \rangle)|}{\sum_o \sum_p (e_{op} - \langle e_o \rangle)^2} \tag{2}$$

where h_p is activation of the candidate hidden unit for pattern p, $<h>$ is the mean activation of the candidate hidden unit for all patterns, e_{op} is residual error at output o for pattern p, and $<e_o>$ is mean residual error at output o for all patterns. C is the absolute covariance between hidden-unit activation and network error summed across training patterns and output units and standardized by the sum of squared error deviations. The same Quickprop algorithm used for output-phase training is employed, but now with the goal of maximizing these covariances rather than reducing error. When the covariances stop increasing, the candidate with the highest absolute covariance is installed into the network, with its just-trained input weights frozen, and a random set of output weights, each with the negative of the sign of the covariance. The other candidates are discarded, thus implementing a realistic proliferate-and-prune cycle.

The basic idea of the input phase is to select a candidate hidden unit whose activations track network error. Once a new recruit is installed, CC returns to output phase to resume training of weights entering the output units in order to determine how to best use the new recruit in error reduction. CC networks have a deep topology with each hidden unit occupying its own layer.

2.2. Sibling-descendant cascade-correlation

A variant called sibling-descendant CC (SDCC) dynamically decides whether to install each new recruit on the current highest layer of hidden units (as a sibling) or on its own new layer, as a descendant.[9] This provides a greedy solution to the dilemma of whether to use flat or deep network topologies.[13] It enables SDCC to create a wider variety of network topologies, normally with less depth and fewer cascaded connection weights. Functionally, SDCC performs much the same as CC on simulations of human experiments.[14]

2.3. *Algorithm extension for learning cessation*

As noted, the SDCC algorithm assesses progress within two phases[9]: an input phase for adjusting input-side weights to recruit a new hidden unit whose activations covary best with network error, and an output phase for adjusting output weights to reduce network error. Each of these phases uses patience and threshold parameters, such that a phase ends when there is no change (whether in error reduction or in correlation with error) greater than threshold for patience epochs. We defined a learning cycle as an input phase followed by an output phase. (The very first learning cycle has an output phase, but no input phase.) To assess progress across learning cycles, we implemented an analogous outside loop to assess progress at the end of each learning cycle, according to the following algorithm, in which a counter is initialized to 0 [3]:

If first learning cycle, record current error and continue to input phase

```
Otherwise, compare current error to previous error as absolute difference
    If absolute error difference > threshold x previous error,
        then reset counter to 0 and continue to input phase
    Otherwise,
        If counter = patience, then abandon learning
        Otherwise, increment counter by 1 and continue to input phase
```

2.4. *The continuous XOR problem*

We tested the extended algorithm on a continuous version of the exclusive-or problem, in which the simplicity of binary XOR is replaced by a more complex continuous version.[15] Starting at 0.1, input values are incremented in steps of 0.1 up to 1, producing 100 x, y input pairs partitioned into four quadrants of the input space (see Figure 1). There is only one output unit, with a sigmoid activation function. Target values of x up to 0.5 combined with values of y above 0.5 yield a positive output target (0.5), as do values of x above 0.5 combined with values of y up to 0.5. Input pairs in the other two quadrants produce a negative output target (-0.5).

We chose this problem because it is well understood[15,16] and easily enables manipulation of varying degrees of difficulty. It also is abstract enough to cover the tasks in various experiments with human infants.

2.5. *The experiment*

To implement problems of differing difficulty levels, we varied learnability, here defined as percentage of target outputs that are not randomly selected: 0, 50, 75, 80, 85, 90, 95, or 100. If a fresh random number in the range [1, 99] ≥ the percent learnability, then the target output (-0.5 or 0.5) is selected with a .5 chance. Patience was 2, 4, or 8, while threshold was 0.1, 0.15, 0.2, 0.25, or 0.3. We ran 20 SDCC networks in each of these 120 cells of the experiment. Each network learned until it reached victory (with correct responses on each training pattern, within a score-threshold of 0.4), or failed to make progress, as defined by the cessation algorithm.

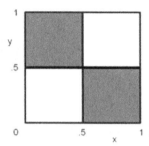

Fig. 1. The continuous-XOR problem, in which gray sectors yield a positive output and white sectors a negative output.

3. Results

There were two dependent variables: whether or not a network achieved victory, and the number of learning cycles it covered. Each was subjected to a full factorial ANOVA with learnability, patience, and threshold as the factors. Only the largest significant effects are presented here. For victories and learning time, we present the main effects in double-y plots of patience ($F(2, 2280) = 224$ for victories, 735 for time), threshold ($F(4, 2280) = 156$ for victories, 283 for time), and learnability ($F(7, 2280) = 587$ for victories, 475 for time), each $p < .001$. Figures indicate both means and standard-error bars so that the significance of mean differences can be assessed at a glance.

Figure 2 shows that victories and learning time both increase with patience. Figure 3 shows that victories and learning time decrease with increasing threshold. Figure 4 shows that victories continually increase with learnability, while learning time is a curvilinear function of learnability thus exhibiting the Goldilocks effect noted in infants.

Note that the x-axis is irregular in these plots, in the sense of not including all values of x with a constant increment.

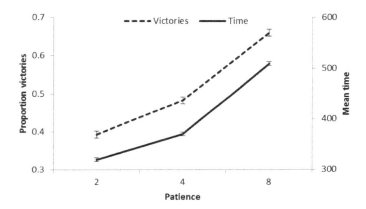

Fig. 2. Proportion of learning victories and mean learning time increase with patience.

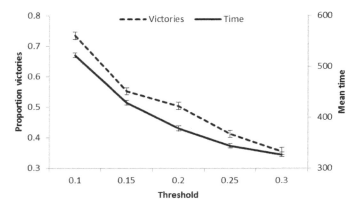

Fig. 3. Proportion of learning victories and mean learning time decrease with increasing threshold.

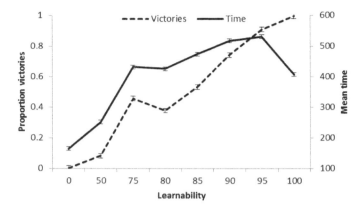

Fig. 4. Proportion of victories and mean learning time as a function of learnability.

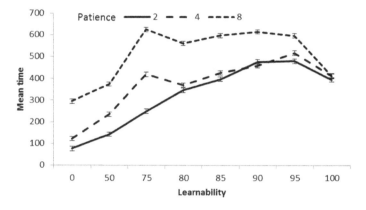

Fig. 5. Mean learning time is a curvilinear function of learnability at each level of patience.

For learning time, there were also significant interactions between learnability and both patience ($F(14, 2280) = 30$) and threshold ($F(28, 2280) = 25$), each with $p < .001$. There is a Goldilocks effect at each level of patience (Figure 5), and each level of threshold (Figure 6). Notably, the time peak is more to the left (low learnability) as patience increases, and to the right (high learnability) as threshold increases.

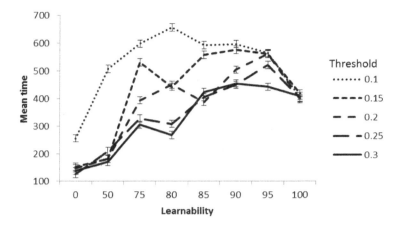

Fig. 6. Mean learning time is a curvilinear function of learnability at each level of threshold.

4. Discussion

Our basic results can be summarized as follows. Random patterns increase learning difficulty, which in turn lowers success and has a curvilinear (inverted U) relation to learning time. This Goldilocks effect is modulated by both the patience and threshold parameters. Increased patience extends learning time, while increased threshold shortens learning time by making the learner less sensitive to changes in error. Although the Goldilocks effect is rather ubiquitous, occurring at every level of patience and threshold tested, the location of its peak on problems of differing difficulty levels varies with patience and threshold. The Goldilocks peak is found at a lower learnability as patience increases and at higher learnability when threshold increases. This relation even becomes linear as the decrease in learning time disappears at extremely high threshold or extremely low patience. Essentially, the model enables predictions of when to expect a Goldilocks curve or just a linear change in learning time, as learnability varies.

Because of their opposite effects on learning time, patience and threshold can compensate for each other. If considered in the context of extending learning time on a learnable task, decreasing threshold can compensate for too little patience, and increasing patience can compensate for a threshold that is too high. This understanding could be important in designing educational interventions.

Our model has some similarity to other autonomous learning systems, particularly in the need to self-monitor learning progress. However, it is based on a small modification to algorithms (CC and SDCC) that cover many phenomena in learning and development[17], without requiring the complexities built into multi-modular and multi-level systems. Our approach is apparently unique in its use of patience and threshold parameters, enabling coverage of infant experiments and novel predictions.

The ability of our model to naturally avoid over-fitting of training patterns, without resorting to programming hacks like validation test sets, enhances its utility for both simulation of natural learning and applications to robotics.

Because work on autonomous learning is relatively new, this and other systems are still far short of complete autonomy. We believe, however, that our system can be extended to cover example selection, multi-task learning, and other important phenomena. Also, costs and benefits could be quantified to enable more precise predictions.

Acknowledgements

This research was supported by a Discovery grant to TRS from the Natural Sciences and Engineering Research Council of Canada. We are grateful to Vincent Berthiaume for relevant pilot research and to a number of colleagues for helpful comments at various stages of the work: Nick Chater, Scott Fahlman, LouAnn Gerken, Caitlin Mouri, and Simon Reader.

References

1. S. E. Fahlman and C. Lebiere, in *Advances in neural information processing systems 2* edited by D. S. Touretzky (Morgan Kaufmann, Los Altos, CA, 1990), pp. 524.
2. T. R. Shultz and S. E. Fahlman, in *Encyclopedia of Machine Learning, Part 4/C*, edited by C. Sammut and G. I. Webb (Springer-Verlag, Heidelberg, Germany, 2010), pp. 139.
3. T. R. Shultz, E. Doty, and F. Dandurand, in *Proceedings of the 34th Annual Conference of the Cognitive Science Society*, edited by N. Miyake, D. Peebles, and R. P. Cooper (Cognitive Science Society, Austin, TX, 2012), pp. 2327.
4. J. Schmidhuber, IEEE Transactions on Autonomous Mental Development (1990-2010) **2** (3), 230 (2010); J. Schmidhuber, in *Developmental Robotics 2005 AAAI Spring Symposium* (Stanford University, CA, 2005).
5. A. Baranes and P. Y. Oudeyer, *IEEE Transactions on Autonomous Mental Development* **1**, 155 (2009)

6. P. Y. Oudeyer, F. Kaplan and V. Hafner, *IEEE Transactions on Evolutionary Computation* **11**, 265 (2007).

7. C. Kidd, S. T. Piantadosi, and R. N. Aslin, in *Proceedings of the 32nd Annual Conference of the Cognitive Science Society*, edited by S. Ohlsson and R. Catrambone (Cognitive Science Society, Austin, TX, 2010), pp. 2476; C. Kidd, S. T. Piantadosi, and R. N. Aslin, PLoS ONE 7, e36399 (2012).

8. Lou Ann Gerken, Frances K. Balcomb, and Juliet L. Minton, *Developmental Science* **14**, 972 (2011).

9. S. Baluja and S. E. Fahlman, *Report No. Technical Report CMU-CS-94-209*, (1994).

10. M. C. Nelson and W. T. Illingworth, *A practical guide to neural nets.* (Addison-Wesley, Reading, MA, 1991); L. Prechelt, in *Neural networks: Tricks of the trade*, edited by G. B. Orr and K.-R. Mueller (Springer, Berlin, 1998), pp. 55

11. C. Wang, S. S. Venkatesh and J. S. Judd, *Advances in Neural Information Processing Systems* **6**, 303 (1994).

12. S. E. Fahlman, in *Proceedings of the 1988 Connectionist Models Summer School* edited by D. S. Touretzky, G. E. Hinton, and T. J. Sejnowski (Morgan Kaufmann, Los Altos, CA, 1988), pp. 38.

13. F. Dandurand, V. Berthiaume, and T. R. Shultz, *Connection Science* **19**, 223 (2007).

14. T. R. Shultz, in *Processes of change in brain and cognitive development: Attention and performance XXI.*, edited by Y. Munakata and M. H. Johnson (Oxford University Press, Oxford, UK, 2006), pp. 61.

15. T. R. Shultz and J. L. Elman, in *Advances in Neural Information Processing Systems 6*, edited by J. D. Cowan, G. Tesauro, and J. Alspector (Morgan Kaufmannn, San Francisco, CA, 1994), pp. 1117.

16. T. R. Shultz, Y. Oshima-Takane, and Y. Takane, in *Advances in Neural Information Processing Systems 7*, edited by D. Touretzky, G. Tesauro, and T. K. Leen (MIT Press, Cambridge, MA, 1995), pp. 601.

17. T. R. Shultz, *Cognitive Development* **27**, 383 (2012).

A CONFLICT/CONTROL-LOOP HYPOTHESIS OF HEMISPHERIC BRAIN RESERVE CAPACITY

N. RENDELL and E. J. DAVELAAR

Department of Psychological Sciences, Birkbeck, University of London
Malet Street, London, WC1E 7HX, United Kingdom

A computational model is presented which draws upon previous research on prefrontally mediated executive functioning to investigate resource allocation in response to increased task demand. Central to the model is the process of monitoring task demand through detection of cognitive conflict. The model is shown to account for performance in a stimulus matching task in which participants are matching stimuli that are presented in the same or opposite visual hemi-field. The model captures the within-hemispheric advantage in both young and older adults and the between-hemisphere advantage when the cognitive demand increases. We conclude that this model provides a first step in the direction of unpacking the mechanisms underlying cognitive and brain reserve capacity.

1. Introduction

Aging affects a number of cognitive mechanisms[1,2]. However, the impact of decline may be offset by further mechanisms which act in response to increased task demand[3,4]. The Cognitive Reserve (CR) hypothesis[4,5] provides an explanation of potential moderating factors that exist between age and cognitive decline. *Passive reserve* refers to a threshold model of resilience (Stern, 2002) in which individuals possess differing amounts of *Brain Reserve Capacity* (BRC). Different amounts of BRC results in differing amounts of protection against brain damage before a threshold is reached and symptoms are manifested at a clinical level. *Active reserve* represents a different, but not mutually exclusive, construct. Rather than an excess of biological substrate, active reserve applies to individual differences in the variability of brain networks and cognitive processes that have become more *efficient* through repeated use[5]. Specific to this study, active reserve may also refer to the concept of neural compensation which involves the recruitment of brain areas not normally associated with a particular task[4]. In support of this final proposition, over-activation in additional brain regions of healthy older adults has been observed in response to increased task demand (for a review see Reuter-Lorenz & Park[6]).

The model presented in this chapter focuses on the interaction between two specific brain regions, the rostral prefrontal cortex (rostral PFC) and the anterior cingulate cortex (ACC). We propose that in tasks that would normally demonstrate strong lateralised brain activation, these areas are implicated in the recruitment of contralateral brain regions to combat increased task demands. In terms of our model, the rostral PFC is part of a control network which is activated when stimulus level conflict is detected by a monitoring system represented by the ACC.

Conflict is often associated with the presence of competing responses. A classic example is the Stroop test[7] in which participants are required to name the colour of a series of words denoting colours which are either printed in the colour that they denote or a different (incongruent) colour. Longer response times for incongruent stimuli are thought to arise from the competition between the more automatic word-reading and the required colour naming (reading the word would be the prepotent response). Response conflict, such as this, elicits a robust response from the ACC[8]. However, conflict can occur on a number of processing levels[9]. The most relevant to this study is stimulus conflict. This occurs at the stimulus encoding level of processing when a number of stimuli are presented, only some of which are task relevant[10]. For this study, we are focusing on stimulus conflict since it best reflects the increase in stimuli present in the harder conditions of the behavioural test under consideration.

Previous computational models of hemispheric processing have explored the benefits of divided processing, the majority of which have focused on the complementary aspects of differing connectivity between hemispheres (e.g. Jacobs & Kosslyn[11]; Levitan & Reggia[12]). However, a neural network model by Monaghan & Pollman[13] specifically explores the relationship between bilateral recruitment and task demand. The Bilateral Distribution Advantage (BDA) describes the observed increase in performance when a task is presented to both hemifields compared to a single hemifield. The model was trained to carry out name and shape matching of two letters presented both unilaterally and bilaterally to a network with two adjoined hidden layers representing the hemispheres. Training itself was biased towards the amount of matches present since learning would not occur given the preponderance of nonmatched items in the training set. The results demonstrated an interaction between task difficulty (as represented by task type) and presentation (either unilateral or bilateral) with the advantage being for bilateral presentation of more complex stimuli.

The model behaved in this manner due to the easy mapping in the shape matching task which is nearly complete after two timesteps, whereas the letter matching task took more than two timesteps to complete. Simultaneously, the activation from one hemisphere is transferred to the other hemisphere from timestep two onwards. Thus, in the shape matching task, the model already responds before the activation is transferred to the ipsilateral hemisphere. Presenting the stimulus over both hemispheres will delay the process and thus a bilateral disadvantage is observed. For the letter matching task, the model requires more processing time. By presenting the letters over both hemispheres, the bilateral processing will start sooner leading to a BDA for complex but not simple tasks. Here, we are interested in the *dynamic control* of the hemispheric connection in response to task demand. Thus, we aim to provide a dynamic explanation of the *active recruitment of resources in response to task load*.

Computational models of specific rostral PFC function are not prevalent in the literature. However, computational models of the PFC in general represent this region as the seat of cognitive control. As such, the PFC enables processing to be carried out in accordance within the rules that constrain the current task being performed. For example, in a neural network model of the Stroop task[14], the PFC is represented by an additional (context) layer which acts to bias lower-level information transfer, thus controlling the execution of behaviours which may be more compelling but not relevant to task completion. Models have also been produced of the PFC and its interaction with other systems. For example, the role of dopamine in the PFC as a stabiliser and/or neuromodulator in working memory processes has been investigated using neural network models (for a review see Cohen, Braver, & Brown[15]). Furthermore, PFC interaction with the basal ganglia was represented in a related neural network model[16] to demonstrate how the two subsystems can give rise to a more *selective gating* mechanism to facilitate efficient working memory updating.

Other models of the PFC have implicated the anterior cingulate cortex (ACC) in the role of conflict monitoring and feedback. This would then regulate the application of PFC-mediated cognitive control[17,18]. For our purposes here, it is important to highlight that increased ACC activation is observed in response to increased task demand[4]. Therefore, the combination of the rostral PFC and the ACC provide the functional component of the current research.

The rostral PFC, as a specific region of the PFC, demonstrates activation over a large number of cognitive tasks such as memory, problem solving and perception[18]. However, lesions to this region do not affect performance unless the tasks are not explicit in the course of action to take or when self-sustained maintenance is required [19-21]. The Gateway hypothesis[19-21] was conceived to provide an explanation for the function of this area. This hypothesis holds that that processing carried out in the rostral PFC is crucial to a Supervisory Attentional Gateway (SAG). The SAG is a cognitive system which deals with attending to either stimulus-orientated or stimulus-independent processing. Rather than contributing to processing within a specific modality, the rostral PFC acts as a switching mechanism that redirects processing flow from externally generated stimulus to self-generated thought. Therefore, the distinction is made between stimulus-independent thought (SIT) and stimulus-orientated thought (SOT). Furthermore, studies such as Gilbert, Frith, & Burgess[22] provide additional evidence for a functional organisation which supports the hypothesis. Three separate tasks, each of which cycled between reliance upon stimulus-orientated and stimulus-independent thought were presented to participants in a fMRI study. The results indicated that the part of the tasks that required the participant to respond directly to the visually presented information, thus requiring stimulus-orientated cognition, demonstrated sustained activation in the *medial* rostral PFC. Task phases requiring just stimulus-independent thought did not demonstrate this sustained activation. Furthermore, activation in the *lateral* rostral PFC occurred only temporarily during the period of switching between the two task phases.

We propose a computational account of the processes involved in demand driven inter-hemispheric communication. Our account is based on two testable assumptions:

1) increased cognitive demand leads to increased cognitive conflict
2) increased stimulus level cognitive conflict is used to upregulate inter-hemispheric communication.

The first assumption requires a neural system that monitors cognitive conflict[18]. This neural system involves the anterior cingulate cortex (ACC) (and possibly other medial prefrontal areas). The second assumption is more speculative and requires a neural mechanism that influences inter-hemispheric communication. For this, we propose that in addition to the function of process flow control between internal and external events, the functionality of the rostral PFC extends to include pathway control between single and dual hemisphere processing.

Together, these assumptions are sufficient to observe a bi-hemispheric neural network that recruits additional neural resources in the face of increased task demand. We include a simulation of age effects, as underlying the HAROLD model[23] and the work of Reuter-Lorenz et al.[3].

The model builds on previous work on the activation buffer[24,25] and extends this by including a conflict/control loop[26].

2. Methods

2.1. *Behavioural data*

The task used in this study was taken from a visual field study by Reuter-Lorenz *et al*.[3] The task used a visual field paradigm in which separate stimulus letters were projected simultaneously to each visual hemi-field (see Figure 1). A probe was also concurrently displayed in one of the hemi-fields. The task included trials that provided matches between stimulus and probes that occurred either between or within hemi-fields. The response required was a go/no-go decision in which the participant stated whether a match was present between any of the stimulus letters and the probe. Task demand was manipulated through stimulus difficulty at three levels. Performance on both conditions was also compared between older and younger adults. Reuter-Lorenz *et al*. found that at higher levels of task demand, the older participants demonstrated an advantage for inter-hemispheric processing.

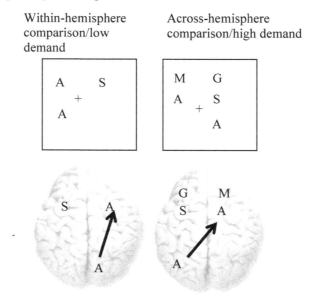

Fig. 1. Schematic of within-hemisphere and across-hemisphere trials with examples of low and high demand stimuli, taken from Reuter-Lorenz et al. 1999. Within-hemisphere matches are represented by the correct target and probe in the same hemisphere. In between-hemisphere matches, the location of the correct match requires across hemisphere processing.

One of the suggestions for the pattern of results observed in older adults is the result of a young brain working harder[27]. In the context of the model presented in this study, this point of view suggests a general advantage for bi-hemispheric processing when the task demand exceeds the resources provided by unilateral processing. Specific to this model, the increase in inter-hemispheric activity is due to control processes which result from increased buffer activation due to an increase in task demand.

2.2. Simulation method

The model is a dynamical system in which activation produced by the system acts as a proxy for response time for the purpose of this investigation. Therefore, the conditions for which activation is greatest can be considered to reflect the fastest response time in that task. The model comprises of three components (see Figure 2). The first component is a localist representation of each of the task outcomes given letters 'A', 'S', 'M' & 'G' presented over both hemispheres with a matching probe (in this case 'A') presented to one hemisphere or the other. Therefore the input comprises ten 'nodes', left and right representations of A-G plus probe for each hemisphere. The inputs were connected to a second component, an activation based buffer via weighted connections between each of the local representations and each item in the buffer. The initial weights represented both strong (1.0) within-hemisphere connections and weak (0.5) between-hemisphere connections.

The buffer in our model was capacity limited[24,25]. This was achieved by lateral inhibitory connections providing competition among items. The resulting capacity limitation was due to items dropping from the buffer when too many are active at the same time (see Davelaar et al.[24] for a full description of the capacity limited activation buffer). Each item in the buffer also has self-excitatory connections to enable activation after the initial presentation of the stimulus.

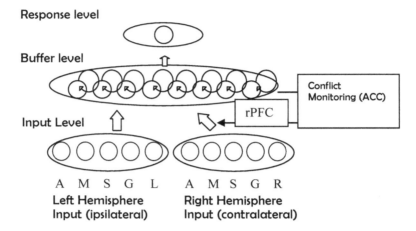

Fig. 2. Architecture of the resource flow model. The model consists of an input connected to the buffer via weighted connections. A conflict monitor (ACC) feeds into the rostral PFC which adjusts the size (weight) of the connection between hemispheres according to the amount of conflict between items in the buffer.

The central roles of the PFC in process flow control and the conflict monitoring role of the ACC was represented by the third component of the model which was the weight adjustment of the input connected to the contralateral hemisphere in response to the increase in stimulus level conflict. The conflict was monitored at the level of lateral connections between items in the capacity limited buffer. This has been successfully applied to related ideas such as producing confidence ratings[26] and stimulus conflict in the flanker task[28]. Therefore, in our model, an increase in stimulus load would increase the number of buffer items which would in turn increase the inhibitory activation in the lateral connections. This would result in an increase in the influence of the contra-lateral hemisphere.

A number of parameters dictate the activity level χ. These factors are represented in Equation 1, where α is the strength of the self-recurrent connection of each item. β relates to strength of the lateral inhibitory connections to the other items and λ is the Euler integration constant.

$$\chi_i(t+1) = \lambda\chi_i(t) + (1-\lambda)[\alpha_i F(\chi_i(t))$$

$$-\beta\sum_{j\neq i}^{N} F(\chi_j(t)) + I_i(t) + \xi_i(t)]....(1)$$

It is also possible that on each time step, the activation is supplemented with zero-mean Gaussian noise with standard deviation σ. However, noiseless models were produced in this investigation. To represent both younger and older adults, α, the strength of self-connectivity of the units in the buffer was adjusted. Older adults were represented by $\alpha=1.6$ and younger adults by $\alpha=2.2$ [24]. The choice of α comes from the interpretation of α as reflecting the neural dynamics of intra- and inter-areal reverberatory loops which become less efficient with age. To demonstrate that the inclusion of a conflict monitoring function contributes to inter-hemispheric processing, models were produced with and without this function. All other parameters remained constant throughout ($\beta = 0.2$, $\lambda = 0.98$).

2.3. *Testing*

The model ran for a total of 1000 time steps for each of four conditions. Each condition was represented by activating the input nodes relating to the both within and between for easy and difficult task demand. Therefore, the four conditions were within-easy, within-hard, between-easy and between-hard. The localist representation of each condition was then multiplied by the weight matrix containing between and within weights at each related position. These activation levels were the starting activity states of items in the buffer. Therefore, easy matches were represented by activation in two buffer items and hard matches were represented by activation in four items. Activation of each of the items was calculated according to Equation 1. At each time step, the weight matrix was adjusted in accordance with the moderating function. Since all cases in this investigation represented a match between probe and stimulus, the settled activation level of the item representing the matched probe was used as a proxy for the speed of the match response.

3. Results

For direct comparison to the behavioural study, activation of within hemispheres was subtracted from activation between hemispheres. Final activation data was recorded for the model both inclusive and exclusive of the conflict monitoring process (See Figures 4 and 5 respectively). For comparison, net response times from Reuter-Lorenz *et al.*[3] are provided by subtracting the within-hemisphere response times from the between-hemisphere response times (see Figure 3).

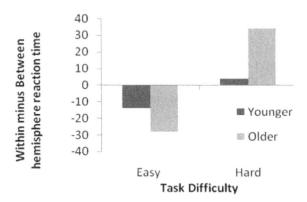

Fig. 3. Results reported by Reuter-Lorenz et al.[3] Note: bars represent within-hemisphere minus between-hemisphere response times: negative scores reflect faster within-hemisphere processing.

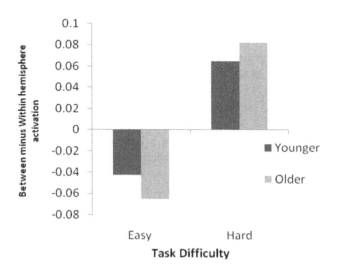

Fig. 4. Results of model *with* conflict monitoring. The difference in hemispheric activation is used as a proxy for differences in cognitive processing as measured with response times.

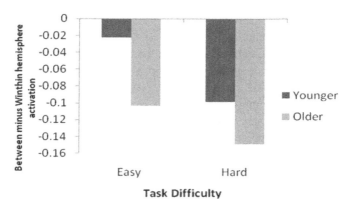

Fig. 5. Results of model *without* conflict monitoring. The difference in hemispheric activation is used as a proxy for differences in cognitive processing as measured with response times.

4. Discussion

Here, we have tested a new mechanistic account of Brain Reserve Capacity through increased hemispheric communication, which has been triggered by increased task demand. The results from this study demonstrate a pattern of activation that reflects the cerebral activation observed in response to a greater cognitive load. In the case of the HAROLD model[23], this is reflected by the relative difficulty, posed by the same task, between younger and older participants. Therefore, this model provides an account of the activation viewed in the HAROLD literature. Furthermore, when increased activation was viewed as a proxy for decreased response time, the pattern of response times for within-hemisphere and between-hemisphere over easy and hard task conditions is similar to that of the behavioural study carried out by Reuter Lorenz et al. [3]. These results provide us with a picture of the processes involved in the bilateral recruitment of resources in the face of increased task demand.

While we presented the model as a proof of concept of demand-based recruitment, the qualitative fit to the data allows us to make some tentative predictions. Primarily, these involve the pattern of data in response to varying levels of task demand beyond those investigated in this study. Although only two levels of task demand from the behavioural study have been used, the trend for faster response time across hemispheres is shown for both medium and higher levels of task demand in Reuter-Lorenz et al.'s[3] study remains the same. Therefore, increasing levels of task demand presented to the model should result in the same pattern at output. However, at a certain level of task demand it is possible that the amount of control that can be exerted reaches its maximum.

Therefore, future directions should include a systematic study of the model over varying levels of task demand.

The results capture the data patterns from both a behavioural study and the imaging literature. Our model was applied to a simple task to highlight the strength of its conceptual underpinnings. It demonstrates that conflict monitoring can play an important part in the recruitment of neural resources. However, it may be necessary to go beyond the scope of descriptive models in order to investigate this process in more detail. The overall outcome of our model was the same as that demonstrated in Monaghan & Pollmans[13] neural network model. We have demonstrated that harder tasks benefit from bilateral processing. However, our model was concerned with how online recruitment takes place as part of active cognitive reserve. One possible manipulation would be to lesion a bilateral network once trained. This would reduce the amount of available resources and, given our theoretical framework, equate to a subjective increase in task demand. Another way of comparing the two models would be to adjust task load with a single task by increasing or decreasing the amount of stimulus available at input. However, the possible limitations of applying this approach to a neural network model have been described in the introduction. This would prove difficult to demonstrate with a neural network model due to similar outcomes for easy and hard tasks and crosstalk between inputs if task load was simply a function of increased stimulus as we have produced in our model.

The use of neural networks may provide a basis for further behavioural and neuroimaging studies. Such work could look at the effect of age-related parameter changes upon resource allocation. For example, age-related changes to the neural substrate underlying either conflict monitoring or the activation-based buffer may explain the individual differences observed when imaging human participants under differing task conditions. This may be carried out by more subtle manipulation than lesioning, such as changing the gain in a sigmoid transfer function to simulate the reduction in dopamine during aging[29]. This manipulation has also been successfully applied to the investigation of dopamine disturbance in schizophrenia and modeling the pharmacological effects of dopamine on learning[14].

One caveat of our study is that demand itself cannot be directly compared between that experienced by human participants and by the model. However, systematic study may reveal relationships in the data common to both the model and human participants, including the expectation of a close functional relationship between the ACC and rPFC during resource allocation.

In this article, we have presented a model of resource allocation in response to increased task demand using conflict monitoring as a trigger for cognitive control. Our proposal builds on a combination of four separate literatures: activation-based working memory, conflict monitoring and control, functional significance of the rostral PFC, and the large literature on cognitive and brain reserve capacity. The model suggests a plausible and testable mechanism in which contra-lateral regions are recruited in task processing when task demand increases relative to the functional capability of the dominant hemisphere.

References

1. K. L. Bopp and P. Verhaeghen, The journals of gerontology. Series B, Psychological sciences and social sciences **60**, 223 (2005).
2. D. C. Park, G. Lautenschlager, T. Hedden, N. S. Davidson, A. D. Smith and P. K. Smith, Psychology and Aging **17**, 299 (2002).
3. P. A. Reuter-Lorenz, L. Staczak and A. C. Miller, Psychological Science **10**, 494 (1999).
4. Y. Stern, Neuropsychologia **47**, 2015 (2009).
5. Y. Stern, Journal of the international neuropsychological society **8**, 448 (2002).
6. P. A. Reuter-Lorenz and D. C. Park, The Journals of Gerontology Series B: Psychological Sciences and Social Sciences **65**, 405 (2010).
7. J. R. Stroop, Journal of Experimental Psychology **18**, 643 (1935).
8. D. M. Barch, T. S. Braver, E. Akbudak, T. Conturo and J. Ollinger, Cerebral Cortex **11**, 837 (2001).
9. V. Van Veen, J. D. Cohen, M. M. Botvinick, V. A. Stenger and C. S. Carter, NeuroImage **14**, 1302 (2001).
10. S. A. Bunge, E. Hazeltine, M. D. Scanlon, A. C. Rosen and J. D. E. Gabrieli, J. D. E., NeuroImage **17**, 1562 (2002).
11. R. A. Jacobs and S. M. Kosslyn, Cognitive Science **18**, 361 (1994).
12. S. Levitan and J. A. Reggia, Neural Computation **12**, 2037 (2000).
13. P. Monaghan and S. Pollmann, Journal of experimental psychology. General **132**, 379 (2003).
14. J. D. Cohen, T. S. Braver and R. C. O'Reilly, Philosophical transactions of the Royal Society of London. Series B, Biological sciences **351**, 1515 (1996).
15. J. D. Cohen, T. S. Braver and J. W. Brown, Current Opinion in Neurobiology **12**, 223 (2002).
16. M. J. Frank, B. Loughry and R. C. O'Reilly, Cognitive, affective & behavioral neuroscience **1**, 137 (2001).
17. M. M. Botvinick, T. S. Braver, D. M. Barch, C. S. Carter and J. D. Cohen, Psychological Review **108**, 624 (2001).

18. M. M. Botvinick, J. D. Cohen and C. S. Carter, Trends in Cognitive Sciences **8**, 539 (2004).
19. P. W. Burgess, I. Dumontheil and S. J. Gilbert, Trends in Cognitive Sciences **11**, 290 (2007).
20. P. W. Burgess, A. Quayle and C. D. Frith, Neuropsychologia **39**, 545 (2001).
21. P. W. Burgess, S. K. Scott and C. D. Frith, Neuropsychologia **41**, 906 (2003).
22. S. J. Gilbert, C. D. Frith and P. W. Burgess, European Journal of Neuroscience **21**, 1423 (2005).
23. R. Cabeza, Psychology and Aging **17**, 85 (2002).
24. E. J. Davelaar, Y. Goshen-Gottstein, A. Ashkenazi, H. J. Haarmann and M. Usher, Psychological Review **112**, 3 (2005).
25. E. J. Davelaar, Adaptive Behavior **15**, 51 (2007).
26. E. J. Davelaar, in J. Mayor Ruh, N., Plunkett, K. (Ed.), Connectionist models of behaviour and cognition II, 241 (2009).
27. P. A. Reuter-Lorenz and K. A. Cappell, Current Directions in Psychological Science **17**, 177 (2008).
28. E. J. Davelaar, Brain Research **1202**, 109 (2008).
29. S.C. Li, U. Lindenberger and S. Sikström, Trends in cognitive sciences **5**, 479 (2001).

Action and Emotion

MODELING THE ACTOR-CRITIC ARCHITECTURE BY COMBINING RECENT WORK IN RESERVOIR COMPUTING AND TEMPORAL DIFFERENCE LEARNING IN COMPLEX ENVIRONMENTS

J. J. RODNY and D. C. NOELLE

Cognitive and Information Sciences.
University of California, Merced

Humans are often able to adapt their behavior, based on experience, so as to optimize the likelihood of obtaining future rewards. These reinforcement learning (RL) phenomena have been well captured by computational models employing temporal difference (TD) learning. TD learning typically involves an "actor-critic architecture" (AC), in which an "adaptive critic" module learns to predict future reward. These models are proven to converge to optimal solutions in simple RL environments; however the proof assumes a discrete environment. The "value function" (VF) in the AC needs to be approximated in continuous and very large discrete environments. Traditional "value function approximators" (VFAs) such as artificial neural networks with back-propagation, while having proven fruitful[1], are not reliable in complex or continuous environments, often times never converging to any solution[2]. Headway has been made in using complex neural network implementations for the VFA; for some problems, convergence to a bounded region of solutions has been proven when using an "echo state network" (ESN), a type of "reservoir computing".[3] More general spiking neural networks have been used to implement an AC architecture.[4] Other work has been done suggesting that "spike-timing-dependent plasticity" (STDP) may be a possible underlying mechanism for computing temporal differences of reward.[5] None of this work has addressed the problem that VFAs in AC architectures fail to learn in a general way in complex environments. We report on simulations involving spiking neural networks including liquid state machines applied to the problem with VFAs in such complex environments demonstrating the benefits and pitfalls using the temporal dynamics of spikes to encode the continuous state value information.

1. Introduction

Humans and other animals, interacting with their environments, are often able to adapt their behavior, based on experience, so as to optimize the likelihood of obtaining future rewards. These *reinforcement learning* (RL) phenomena have been well captured by computational models employing *temporal difference* (TD) learning, in which the difference between two successive evaluations of the organism's state, called the *temporal difference error* (*TD error*), is used to improve both the evaluation process and the process for selecting actions.[1,6,7] Specifically, Barto et al.[8] introduced the *adaptive actor-critic architecture*, in which an *adaptive critic* module employs a TD learning strategy to learn the relationship between the organism's state and the utility, or value, of being in that state (*value function*) in conjunction with an *actor* module learning the relationship between the organism's state and the best action to take in that state (*action policy*). Samuel[6] proved that this approach is guaranteed to converge to an optimal action policy when the learning agent employs a discrete look-up table to maintain its value function. Importantly, there is a growing body of biological evidence for a learning mechanism of this kind in the brain. Specifically, phasic firing patterns of dopamine cells in the basal ganglia appear to encode the temporal difference error, driving synaptic plasticity in a manner matching the learning mechanisms in these reinforcement learning models.[8,9,10]

While TD learning has been proven to converge to the optimal action policy when the learner's value function is implemented as a look-up table from states to values[7], this is not a viable mechanism for learning in the brain. The range of possible situations in which a typical organism may find itself is extremely large, with each state being essentially unique in one way or another. Given a finite amount of past experience, the organism must learn to estimate the value of states that have never been previously experienced, and it must choose appropriate actions in these states. This generalization problem is typically addressed by replacing the discrete look-up table implementation of the value function with a *value function approximator* (VFA): a parameterized function that maps features of the current state to an estimate of its value. Since the domain of this mapping involves features of a state, rather than discrete states, a VFA can produce assessments of novel combinations of features, encoding novel states. Learning, in this case, involves adapting the parameters of the VFA so as to improve the resulting estimates. Indeed, neural networks make for natural VFA implementations.

Unfortunately, guarantees of convergence to optimal action policies do not exist when the value function is implemented by a VFA. Still, this approach has been practically successful in models. For example, others[1,11] have extended Sutton's TD learning methods to learning agents in a context-dependent complex environment, namely backgammon, and found positive (champion level) results using a backpropagation neural network[12] as a VFA. Similarly, Rumelhart et al.[13] used neural networks to approximate the value function and showed that the adaptive critic in the actor-critic architecture functions very similarly to dopamine neurons projecting to the striatum and frontal cortex.

This approach has not been consistently successful, however. Boyan et al.[2] applied TD learning to a variety of problems involving continuous state spaces.

They found that using a VFA for the value function in these problems often resulted in a failure to learn any reasonable action policy.

Sutton[14] replied to Boyan et al.[2] with a modified learning architecture that employed a sparse coarse-coded function approximator that discretized the state space into multiple overlapping tilings, encoding the state within a continuous state space with high precision, while still capturing the similarity structure of the state space in terms of shared tile features. The problem with this solution, however, was that the state representation needed to be hand-coded for each new environment the learning algorithm encountered; thus Sutton[14] simply hard-coded a solution to the problem, rather than providing a new general solution for value function approximators in continuous environments.

Thrun and Schwartz[15] and Smart and Kaebling[16] recognized these problems with VFAs: what they call *hidden extrapolation*[16] or *systematic overestimation*[15] problems. When VFAs extrapolate new data points instead of interpolating them, errors exponentially increase. Smart's response to this problem was to lead the agent through a subset of "interesting" states in the state space within which, the value function approximator could only interpolate new data points, avoiding the problem of *hidden extrapolation*.[15] Note that this solution does not directly address the failure of VFAs, but suggests an alternative training system to compensate for their limitations, instead.

In this article, we report some preliminary work focused on addressing these learning limitations by making use of a novel class of VFA mechanisms. In particular, we investigate the use of a reservoir computing neural network framework called *liquid state machines* as a means for implementing value functions for TD learning in complex environments.

1.1. *Liquid State Machines for Value Function Approximation*

The learning problems observed with VFAs may be a function of the specific parameterized functions that have been used as VFAs. Most VFAs have been implemented as neural networks: single and multilayer perceptrons and recurrent multilayer neural networks have garnered mixed results. Is there another type of value function approximator that might be able to extrapolate to novel states in complex environments more accurately?

One possibility is an *echo state network* (ESN), which is a large reservoir of randomly and sparsely connected neurons. ESNs are part of the *reservoir computing* paradigm of neural networks. This type of recurrent neural network traditionally does not do any learning within the reservoir. It is connected to a layer of output neurons whose connection weights from the reservoir network of neurons are trained in a supervised manner (e.g. delta learning). Szita et al.[3] proposed the use of such an ESN as a value function approximator; they showed that, when using an ESN as a VFA, the TD learning algorithm would converge to a solution, whereas previous work[2] showed divergence from any solution (whether optimal or not) with a standard VFA.

Why might an ESN or other reservoir computing neural network succeed as a VFA when other approaches have failed? One possible reason simply involves the large number of neurons that are typically used in a reservoir network. This idea is inspired by Cover's Theorem[17], in which Cover proves that the probability that a random nonlinear problem can be solved by a linear classifier increases as the nonlinear problem is projected non-linearly onto a higher and higher dimensional space. Reservoir computing essentially performs such a high dimensional projection of its state-feature input nonlinearly into a higher dimensional space spatio-temporally, producing an increased probability of success when a linear value function approximator is given the high-dimensional reservoir activity as input. From this, we hypothesize[3] that reservoir neural networks may outperform traditional recurrent neural networks when used as VFAs.

While a large reservoir of neurons (like an ESN) might prove useful as a VFA, combining the memory capacity of reservoir computing with a spike-based neural model might allow the reservoir to contain a much richer memory. An ESN averages over neuron spikes, ignoring the inter-spike-intervals which can provide much more information to the system than a scalar value of the firing rate.[18] In this work, we investigate implementing VFAs using a spike-based reservoir computing neural network called a *liquid state machine (LSM)*.

Kello *et al.*[19] used a liquid state machine, in conjunction with a self tuning algorithm called *critical branching* to allow the LSM to tune itself to its *critical point*. Having the reservoir on the *edge of chaos,* or at the critical point, is "where memory and representational capacity of the liquid were shown to be optimized".[20]

Including a critical branching mechanism and extending the work done by Szita *et al.*[3] to LSMs might provide extra information in the inter-spike-intervals lost by ESNs, producing a VFA with the power to not only interpolate between previously experienced states, but extrapolate to novel situations more accurately.

2. Methods and Results

We simulated an actor-critic architecture implementing a temporal difference (TD) learning algorithm that uses a liquid state machine (LSM) to implement the *value function approximator* (VFA). The state input features are sent through the LSM, and a linear value function approximator uses the activation of neurons in the LSM reservoir, instead of the environmental features directly, to learn to predict future reward. Both the actor and the adaptive critic use the temporal difference (TD) error from the VFA to learn, using standard weight update mechanisms.[14] In this way, the adaptive critic learns the relationship between LSM features and reward, and the actor learns the relationship between these features and the best action to take in the current state. Experiment 1 describes initial exploratory results, modeling only performance of an LSM as a VFA with no learning, while Experiment 2 describes results from modeling the entire actor-critic architecture with the LSM as a VFA, with learning in both the actor and critic.

2.1. *Simulation Experiment 1 Method*

As an initial test of this approach, we modeled only the VFA with the rest of the actor-critic architecture held constant. Specifically, the adaptive critic did not calculate the temporal difference of successive state evaluations, and the action policy of the actor did not change. This is a simplified framework, allowing us to perform an analysis of how well the LSM performs as a VFA. We wanted to see if the features of the environment would be successfully encoded within the LSM, checking if it could perform at least as well as standard VFAs.

We used a moderately difficult learning problem, involving a continuous environment, for this analysis. We created a two-dimensional "puddle world", as examined by Boyan and Moore[2] (see Figure 3B), and we had a simulated agent take a random correlated walk[a] around the state space of the puddle world. We provided the agent with its Cartesian X and Y coordinates in the state space, as well as an instantaneous "reward" value. For this simulation, the "reward" communicated how deep in a puddle the agent is in its current position at each time step. We trained a linear VFA, given the neural state of the LSM as input, to predict this reward value a fixed interval before and after every time step. In this implementation, the *action policy* of the agent is not modified in any way; throughout, it is a correlated random walk. This is in effect an off-line learning strategy: the *action policy* is fixed throughout the learning period.

As the simulated agent randomly traversed the puddle world, its location in the world in Cartesian X and Y coordinates was provided as input to a large LSM (1000 units), and the instantaneous reward experienced by the agent (i.e., puddle depth) was recorded, with the LSM propagating the location-encoding

---

[a] A random correlated walk is similar to a random walk, however the probability of the agent to continue moving in the same direction as last time step is slightly increased, We ran the simulation with three different such conditions: 90% forward, 4.9% left, 4.9% right, 0.2% backwards; 60% forward, 19% left, 19% right, 2% backwards; and a random walk: 25% forward, 25% left, 25% right, 25% backwards.

input spikes throughout the neural units in its reservoir. The input feature neurons were connected to the neurons of the LSM, and the reservoir initially employs critical branching to bring the LSM weights to a critical point, to produce heavy-tailed distributions of spikes and output.[19] The spiking patterns of the reservoir units at each discrete time step were also recorded, and linear regression was used to attempt to predict the reward value of the state of the environment, given the current state of the reservoir of the LSM.

This first simulation was designed to investigate the ability of the LSM to encode the agent's state sufficiently well to allow a linear VFA to predict instantaneous reward. Also, in comparison to VFAs based on ESNs, we hoped to see if the spiking nature of LSMs allowed them to capture more or less information about the value function.

2.2. Experiment 1 Results

Using an LSM as a VFA gave slightly better evaluations of the current state than simple linear regression on the X and Y positions themselves. The LSM postdicted past rewards better than a straight linear regression on the state inputs, and the LSM predicted future rewards as well as the straight linear regression approach.

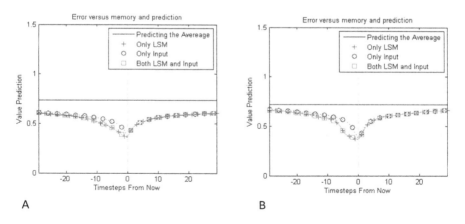

Fig. 1. Reward prediction/postdiction error averaged over each point in time. Runs of the LSM at an uncorrelated random walk (A), and a correlated random walk of (B) 60%. Y axis is VFA error, X axis are time steps in the future and past from the current time step. The center vertical line represents the current time step. The solid line is LSM as a VFA, the circles are the LSM + XY input as VFA, plus signs are the original single layer VFA without LSM, and the squares are the error from predicting the average reward at each time step.

Figures 1 and 2 show the results of the performance of the LSM as a VFA; for each of the graphs, there are four different sets of data used to predict future reward and postdict past reward: the state of the LSM, the current X and Y position, the combination of LSM state and X and Y position, and the average reward for the random walk. The analysis takes the current state of the LSM (or the current XY position, or the combination of the two) and tests how well this predicts the reward at each time step in the future (in the case of figures 1 and 2, for 30 time steps) and past. The error in prediction is averaged over the entire second half of a long walk through the puddle world, and this error is displayed in these first two figures. These graphs show that the LSMs are better than XY position alone at postdicting reward, and they perform no worse than simply using the XY position when predicting future reward.

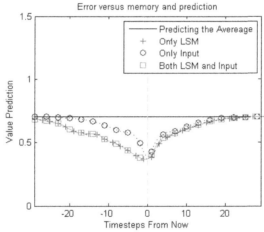

Fig. 2. Reward prediction/postdiction error averaged over each point in time. A run of the LSM at correlated random walk of 90%. Y axis is VFA error, X axis are time steps in the future and past from the current time step.

There is also a difference between Figures 1A, 1B, and Figure 2, in that for the higher structured action policies, especially for the 90% correlated random walk (Figure 2), the LSM improved the postdiction of the VFA substantially versus using only the XY positions. These are robust differences from multiple runs through the simulation. One reason this may be the case is that for random walks, it will be much slower for a truly random walk to leave a puddle, and predicting a similar reward value for each step is an easy solution. However, given an action policy that is highly patterned such that it very consistently follows the same direction (leading to long straight walks and leading to walks straight through a puddle), predicting reward based on the reward last time step will not perform as well. Thus the error for a linear VFA on the XY position, directly, is lower for the uncorrelated random walk case because the reward value doesn't change drastically, which makes reward prediction (and postdiction) much easier.

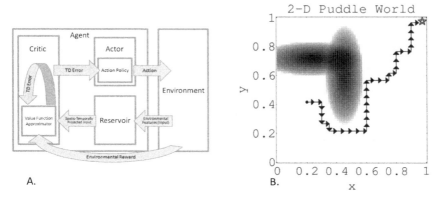

A.   B.

Fig. 3. In the left panel, the actor-critic architecture with an LSM as a VFA. The environment gives the reservoir environmental features, and the critic calculates a value function from the reservoir that will maximally predict environmental reward using the temporal difference (TD) error in reward and prediction for reward. The actor calculates its action policy similarly, using the TD error. In the right panel, puddle world reproduced from.[2] Reward is given to the agent when it reaches the top right corner of this 2d world. The dark circles designate puddles where punishment is given to the agent if it enters that XY position.

2.3. Experiment 2 Methods

In our second simulation experiment, the entire actor-critic architecture was implemented in the puddle world described in[2], with the liquid state machine as a filter on the inputs for the actor and adaptive critic.

The overall TD learning model used was similar to that in Barto[13], but all of the simulated neurons made use of a spiking neural model instead of real-valued activation neurons (because LSMs depend upon neural spiking). The layer of input state feature neurons, which separately encode the current X and Y position of the agent, was similar to that in Experiment 1. These input feature neurons were connected to a large LSM reservoir, like that in Experiment 1, with critical branching bringing the LSM weights to a critical point, as before. The reservoir neurons were connected to a single linear neural unit, which acted as a VFA. The weighted connections to this single linear unit were modified using the delta rule, in the same manner as Barto's[13] TD learning actor-critic architecture, based on the reward signal.

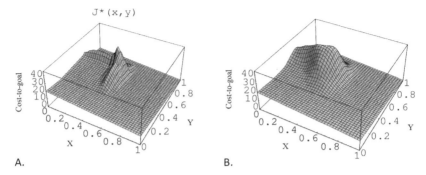

Fig. 4. Reproduced from Boyan and Moore (1995). Left Panel: Optimal VFA output in puddle world as proposed by.[2] Right Panel: Solution to puddle world from Boyan and Moore[2] using an off-line grow-support method not described here.

2.4. *Experiment 2 Results*

Using an LSM as a filter for the input to the VFA within the actor-critic architecture produced promising results. The augmented actor-critic architecture gives a near optimal solution to the puddle world problem (Figure 5). These are similar results to those provided by Boyan and Moore[2] (Figure 4) and Sutton[14], while still learning on-line and without giving the agent a hand-coded representation of the conjunctions of X and Y location features.

While these are promising results, we do find that the value functions learned by this model and the value functions learned by past models do differ, somewhat. First of all, the values output by the adaptive critic in both Boyan and Moore[2] and Sutton[14] capture the cost-to-goal, or the sum of the reward and punishment that the agent receives starting at each position and following the learned action policy. This is easy to calculate with these models, as there is no time dependence of these models on the input. The LSM, however, pulls the location input apart spatio-temporally, and thus the LSM contains many spikes that encode information about location inputs from the past. Therefore, our value function does not evaluate location states, exactly, but only those states as they appear as an integral part of a temporal sequence of states visited by the learning agent. Thus, we display the LSM-based value function in Figure 5 as the mean of the secondary reward (VFA output) at each XY position, averaged over all of the visits made by the agent to that position. The display of this value function reliably shows the puddles as having a negative value and the rewarding corner as having a positive value.

While these results require further analyses, we can conclude that the VFA augmented with an LSM allowed the adaptive critic to converge to a near-optimal solution in a problem where standard VFAs do not converge to *any* solution.

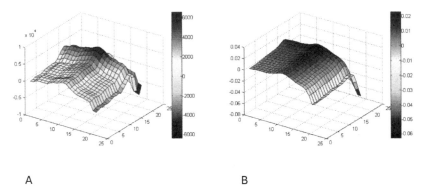

A   B

Fig. 5. Left Panel: The resulting inverse of the secondary reward (the predicted reward values) from the value function approximator (VFA) at each position in the puddle world, where the actor-critic incorporates a liquid state machine (LSM) as a filter on the VFA for a single sample run. Right Panel: The average of the normalized inverse predicted reward from the VFA at each position in the puddle world over 40 independent runs. Note: The inverses of the secondary reward are shown to allow comparison between these and the graphs of the cost-to-goal graphs given in Figure 4, as the secondary reward is a value of how good each position is in getting the agent reward (which is only in the corner), whereas the cost-to-goal is an opposite value to this; how much punishment the agent expects to receive before it reaches the goal.

3. Discussion

We modeled the actor-critic architecture in the puddle world described by Boyan and Moore[2], modifying the actor-critic by implementing a liquid state machine (LSM) as a filter on the environmental features received by the linear value function approximator (VFA) within the agent. Building on past research showing that reservoir computing allows VFAs to converge to a solution (where standard neural network VFAs, such as backpropagation networks, do not)[3], we provide evidence that an LSM can provide a VFA with the ability to solve nonlinear value function learning problems, such as that present in the puddle world paradigm.

Of the many possible avenues for future work, one interesting possibility involves allowing the LSM to adapt to the distribution of its inputs. In essence, this would allow the LSM to conduct unsupervised learning on its inputs in hopes of producing a more useful encoding of the state of the agent. Spike-timing-dependent plasticity (STDP) is a good candidate for driving such unsupervised learning in the spiking neurons of an LSM, as STDP has been shown to be useful in predicting the arrival of a general reward signal in a manner similar to the behavior of the adaptive critic in TD learning.[21] Similar studies by Florian[22,23], Roberts et al.[5], Di Castro et al.[24], and Potjans et al.[4] demonstrated various versions of TD-like reinforcement learning algorithms for spiking neural networks using STDP modulated by a dopamine-like reward signal. Using a synaptic plasticity mechanism similar to these may empower the LSM to pull out useful patterns more easily, facilitating learning by the VFA.

References

1. G. Tesauro, *Machine Learning*, 257 (1992).
2. J. A. Boyan and A. W. Moore, *Advances in Neural Information Processing Systems,* 369 (1995).
3. I. Szita, V. Gyenes, V. and A Lorinez, *Artificial Neural Networks*, 830 (2006).
4. W. Potjans, A. Morrison and M. Diesmann, *Neural Computation*, 301 (2009).
5. P. D. Roberts, R. A. Santiago and G Lafferriere, *Biological Cybernetics*, 517 (2008).
6. A. L. Samuel, *IBM Journal on Research and Development*, 210 (1959).
7. R. S. Sutton, *Machine Learning*, 9 (1988).
8. A. G. Barto, R. S Sutton and C. W. Anderson, *IEEE Transactions on Systems, Man and Cybernetics*, 835 (1983).
9. P. Montague, P. Dayan and T. A. Sejnowski, *Journal of Neuroscience*, 1936 (1996).
10. W. Schultz, *Journal of Neurophysiology*, 1 (1998).
11. J. A. Boyan, (1992).
12. D. E. Rumelhart, G. E. Hinton, and R. J. Williams, *Nature* **323.6088,** 533 (1986).
13. A. G. Barto, *Models of Information Processing in the Basal Ganglia*, **15,** 215 (1995).
14. R. S. Sutton, *Advances in Neural Information Processing Systems*, 1038 (1996).
15. S. Thrun and A. Schwartz, *Proceedings of the Fourth Connectionist Models Summer School*, 587 (1993).
16. W. D. Smart and L. P. Kaelbling, *Machine Learning*, 903 (2000)
17. T. M. Cover, *Electronic Computers, IEEE Transactions on.* **EC-14**, 326 (1965).
18. F. Rieke, D. Warland, R. D. R. van Steveninck and W. Bialek, *MIT press,* (1999).
19. C. T. Kello, G. D. Brown, R. Ferrer-i-Cancho, J. G. Holden, K. Linkenkaer-Hansen, T. Rhodes, et al., *Trends in Cognitive Science* 223, (2010).

20. C. T. Kello and M. R. Mayberry, *International Joint Conference on Neural Networks: Barcelona*, 1 (2010).
21. R. P. Rao and T. J. Sejnowski, *Neural Computation*, 2221 (2001).
22. R. V. Florian, *Symbolic and Numeric Algorithms for Scientific Computing* (2005).
23. R. V. Florian, *Neural Computation*, 1468 (2007).
24. D. Di Castro, D. Volkinshtein and R. Meir, *Neural Information Processing Systems* (2009).

THE CONCEPTUALISATION OF EMOTION QUALIA: SEMANTIC CLUSTERING OF EMOTIONAL TWEETS

E.Y. BANN

Advanced Emotion Intelligence Research
E-mail: eugene@aeir.co.uk
www.aeir.co.uk

J.J. BRYSON

Department of Computer Science, University of Bath,
Bath, BA2 7AY, United Kingdom
E-mail: J.J.Bryson@bath.ac.uk

A plethora of words are used to describe the spectrum of human emotions, but how many distinct emotions exist, and how do they interact? Over the past few decades, several theories of emotion have been proposed, each founded upon a set of basic emotions, and each supported by an extensive variety of research including studies in facial expression, ethology, neurology and physiology. Here we propose a theory that people transmit their understanding of emotions through the language they use that surrounds mentioned emotion keywords. Using a labelled corpus of over 21,000 tweets, six of the basic emotion sets proposed in existing literature were analysed using Latent Semantic Clustering (LSC) to propose the *distinctiveness* of the semantic meaning attached to the emotional label. We hypothesise that the more distinct language is used to express a certain emotion, then the more distinct the perception (including proprioception) of that emotion is, and thus more basic. This allows us to select the dimensions best representing the entire spectrum of emotion. We find that Ekman's set, arguably the most frequently used for classifying emotions, is the most semantically distinct. Next, taking all analysed (that is, previously proposed) emotion terms into account, we determine the optimal *semantically irreducible* basic emotion set using an iterative LSC algorithm. Our newly-derived set (ACCEPTING, ASHAMED, CONTEMPT, INTERESTED, JOYFUL, PLEASED, SLEEPY, STRESSED) generates a 6.1% increase in distinctiveness over Ekman's set (ANGRY, DISGUSTED, JOYFUL, SAD, SCARED).

Keywords: Basic Emotions, Latent Semantic Clustering, Lexical Analysis, Twitter.

1. Introduction

There are a great variety of words that describe the spectrum of human emotion. Many theories posit the existence of a set of 'basic emotions' that are hardwired into our brain as individual neurological circuits [1-5], and that all other emotions are derived from these 'biological primitives' as either a combination or specific valence of these neural circuits [6]. Recently, however, the notion that emotion is a conceptualised act has been proposed [7], and experimental results have been shown to support this hypothesis [8]. Emotion in this sense can be regarded in the same way as colour, insofar we categorise and communicate discrete colours within the confines of language, even though colour itself is in fact a spectrum of visible light.

The primary objective of this research is to evaluate existing basic emotion sets to discern which contain the most number of emotions expressed in the most distinct language, testing the hypothesis that the more distinct an emotion is (that is, unlike any other emotion), the more distinct the language is used to express the experience of that emotion. *Semantics* refers to the meaning of an expression; in particular, we consider co-occurring words to measure similarities of meaning. We attempt to show such semantic changes in emotion language from a corpus of explicitly expressed emotions extracted from the micro-blogging website Twitter, and evaluate six basic emotion sets on a scale of *semantic distinctiveness*, based on the hypothesis that the more distinct the language used to express a certain emotion, then conceptually (i.e. what we understand that emotion keyword to mean), the more psychologically irreducible that emotion is. The less semantically accurate a set of emotions is, the more similar these emotions are to each other, or in other words, if similar words are used when expressing two different emotions, then these emotions are, in theory, conceptually, and thus psychologically, similar. The secondary objective of this research is to identify a set of basic emotions by identifying the most semantically distinct emotion keywords relative to the underlying semantic features of each expression within the corpus.

2. The Psychology of Emotion

Emotion is that which leads the subject's condition to become so transformed that one's judgement is affected [9], triggered by a subconscious appraisal process regarding an issue of personal value [10]. It is characterised by behavioral, expressive, cognitive, and physiological changes [11] and can be started and executed unconsciously [12]. The desire to experience or not experience an emotion largely determines the contents and focus of consciousness throughout the life span [13].

The above definition of emotion is not a conclusive definition of emotion, but amalgamates many of the important aspects from notable theorists' definitions. Attention is drawn to Aristotle's wording, stating that emotion "is *that* which...", implying that emotion is in fact a type of *quale*, that is, a subjective conscious experience that cannot be communicated, or apprehended by any other means other than direct experience [14]. Qualia refers to subjective 'raw' feelings, for example, the taste of red wine, or the experience of seeing the colour red. Emotion qualia thus refers to the raw feel of an emotion; the actual phenomenon of a particular emotion experienced may actually differ according to each person's perception of that emotion.

The dominant theory of emotion postulates the existence of a small set of hardwired, or 'basic', emotions, and consequently the majority of textual emotion recognition research has been based on such.

ANGER DISGUST FEAR JOY SADNESS SURPRISE

Ekman's basic emotion set [15] (shown above) is arguably the most frequently used within the field of computer science for emotion mining and classification. However, not only do the emotions comprising each basic emotion set vary amongst theorists, they do not always agree the definitions of emotion, thus adding to the confusion in delineating the set of basic emotions, or whether they exist at all. This can be viewed as a problem symptomatic of the vagueness of language, which suggests that there is a general problem about how to talk about the emotion qualia [6]. There are two viewpoints concerning the advocation of basic emotion sets: they are either based on biologically primitive or psychologically irreducible emotions (see Bann [16] for a detailed discussion).

Barrett's work [7] studied the act of conceptualising core affect, in other words, why people attach emotion labels to the experience of emotion qualia. She proposed the hypothesis that emotion is a psychological event constructed from the basic elements core affect and conceptual knowledge. In a study focusing on the conceptualisation of *fear* [8], it was found that neither the presence of accessible emotion concept knowledge nor core affect alone was sufficient to produce the (world-focused) experience of fear. As emotions are constructed from conceptual knowledge about the world, we can see that emotions themselves are concepts that human beings begin learning in infancy and continuously extend and revise throughout life

[8]. This repeated experience of labelling a combination of core affect and the context in which it occurs as an emotion provides "training" in how to recognise and respond to that emotion; in this sense, Barrett described emotions as "simulations". This "skill" of conceptualising core affect as an emotion might be a core aspect of emotional intelligence — in much the same way as conceptual thinking is core to cognitive intelligence — defining how humans deal with their internal state, but more importantly, defining the emotion labels used as a combination of specific experiences. Each person's conceptualisation of their emotion spectrum is thus unique; it is this conceptualisation that we attempt to aggregate and analyse in this research.

Emotions can be expressed in a variety of ways including facial expressions, body language, tone of voice, and the language used in speech and text. This research focuses on the most explicit of these — the language used in communication — with the proposition that how humans communicate to one another can reveal individual conceptualisations of specific emotions, given that the specific emotion keyword is used within the communication. Defining *basic emotions* as emotions that are conceptually distinct from any other emotion, we explore the hypothesis that the language used in communicating basic emotions should be significantly different for each one, as each basic emotion should describe a significantly distinct concept.

3. Semantic Analysis

Over the past few decades there has been significant evidence that people's psychological aspects can be predicted through analysis of language style. One notable example is Rosenberg's work on verbal behavior and schizophrenia [17]. He found that, while the speech of those diagnosed with schizophrenia did not differ from unaffected people on the structural level, it did differ on the semantic level, i.e. with regard to the thematic concerns that were being addressed. It is this deviation from expected thematic concerns, which are linked to general and sex-specific social role expectations, that is associated with the diagnosis of schizophrenia [17]. Analysis of language semantics has been used extensively in research, including discovering individual differences in personality [18], lie detection [19] and discovering individual differences in beliefs [20]. With respect to emotion analysis, French found that co-occurrence techniques such as Latent Semantic Analysis does not detect personality from short text samples [21], but do reveal that texts expressing particular emotions have a greater semantic similarity to corresponding exemplar words [22].

By analysing the semantics, specifically, co-occurrence statistics, of the language expressing individual emotion keywords, we can discern those emotions that are similar and those that are distinct. We postulate that similar emotions are represented by similar semantics, and propose to cluster emotional documents based on the underlying meanings of each document.

3.1. *Latent Semantic Analysis*

Latent Semantic Analysis (LSA) [23] is a variant of the vector space model that aims to create a semantic space by means of dimensionality reduction techniques and has been widely used in a variety of domains, from document indexing to essay grading. It has also been used in emotion classification of news headlines, performing better than Naïve Bayes in the case of recall but not as good as WORDNET in terms of precision [24]. Given a raw co-occurrence matrix $\mathbf{M}$ using the entire vocabulary as B, this is transformed by A (the documented function for LSA is log-entropy normalisation), and M is applied to reduce dimensionality.

There are several techniques for M that reduce the dimensionality of words constituting the semantic space, the original method documented for LSA being Partial Singular Value Decomposition (PSVD). PSVD uses Singular Value Decomposition (SVD) to decompose the data matrix $\mathbf{M}$ into the product of three matrices:

$$\mathbf{M} = \mathbf{T}\Sigma\mathbf{D}^T \tag{1}$$

where $\mathbf{T}$ is the term matrix, $\mathbf{D}$ is the document matrix and Σ is a diagonal matrix with singular values sorted in decreasing order that act as scaling factors that identify the varience in each dimension. LSA uses a truncated SVD, keeping only the k largest singular values in Σ and their associated vectors:

$$\mathbf{M} \approx \mathbf{M}_k = \mathbf{T}_k \Sigma_k \mathbf{D}_k^T \tag{2}$$

This reduced-dimension SVD, or PSVD, $\mathbf{M}_k$, is the best approximation to $\mathbf{M}$ with k parameters, and is what LSA uses for its semantic space. The rows in $\mathbf{D}_k$ are the document vectors and the rows in $\mathbf{T}_k$ are the term vectors in LSA space.

4. The Emotional Twitter Corpus

Twitter is a public micro-blogging system that allows users to share short messages of up to 140 characters and, as these are publicly available, provides us with an ethical way of collecting a diverse range of public expressions. Coupled with the fact that a good proportion of tweets project the user's emotion — indeed, French found that some emotions, particularly those with strongly marked valence, can be accurately expressed and perceived in short blog excerpts [25] — we are able to assume that Twitter is a valid sample of human emotive expression and thus a suitable corpus for this project. There is somewhat of an explicit impulse to communicate emotions on Twitter and although the underlying cause is not always explicitly mentioned, it is this factor that we attempt to capture. Our experiences tested a collection of six basic emotion theories as described in Table 1 (see Bann [16] for details of our selection method).

Table 1 Basic Emotion sets from the most notable Basic Emotion theories that were analysed

Basic Emotion Theory	Identified Basic Emotions
Izard	Anger, Contempt, Disgust, Distress, Fear, Guilt, Interest, Joy, Shame, Surprise
Russell's Categories	Angry, Depressed, Distressed, Excited, Miserable, Pleased, Relaxed, Sleepy
Plutchik	Acceptance, Anger, Anticipation, Disgust, Joy, Fear, Sadness, Surprise
Ekman	Anger, Disgust, Fear, Joy, Sadness, Surprise
Tomkins	Anger, Interest, Contempt, Disgust, Distress, Fear, Joy, Shame, Surprise
Johnson-Laird	Anger, Disgust, Anxiety, Happiness, Sadness

4.1. *Emotion Keywords*

The extraction mechanism and the selection of keywords to be mined from Twitter would form the structure of our eventual emotion corpus. Taking the union of all the emotion sets identified for analysis, we obtained a set of 21 unique emotion keywords, which, theoretically, constitutes the most distinct emotions. We extract unigrams created using the first person grammatical inflection of each keyword, similar to Russell [26], as most tweets will contain this type of inflection: *"I am very **excited** today"* as opposed to *"I am feeling **excitement** today"*. This was chosen as opposed to mining for bigrams, for example *"feeling excited"*, *"feel excited"* and *"felt excited"*, as this resulted in far fewer tweets being returned due to Twitter's indexing focusing on single keywords. Moreover, tweets containing quantifiers would have been ignored if we chose to extract bigrams, for example *"feeling **very** excited"*.

Contrary to Bollen's work [27], we did not require tweets to contain the words *'feel'*, *'I'm'*, *'Im'*, *'am'*, *'being'*, and *'be'*, as an explicit mention of an emotion keyword would be sufficient to describe an experience of that emotion, reinforced by the fact that we will only be mining for the first person grammatical inflection of each keyword. We filtered out re-tweets — minimising duplicates — and negative tweets, because, for example, *'happy'* $\neq$ *'not happy'*; nor can we assume that *'not sad'* = *'happy'*. Tweets containing popular phrases which include *"Happy Birthday"* and *"Angry Birds"* were also filtered out. Initially, @ tags were not filtered, but we quickly realised that these tweets refer to messages either closely relating to other people or as part of a thread of messages; thus we filtered them out as the emotion expressed within such tweets did not describe an atomic emotional experience. We did not Porter stem collected words as Kim[28] notes that this might hide important semantic differences, for example, conceptual differences between *loved* and *loving*. To optimally harvest emotions, we substituted *fear* with *scared* as it was proven to be the most popular keyword out of *scared*, *frightened* and *afraid*. We also substituted *distressed* with *stressed*, due to an extremely low stream rate for this keyword (see Bann[16] for a detailed description of keywords).

4.2. Emotion Streaming

A PHP script was created that used the Gardenhose Level Twitter Streaming API — a streaming sample of about 10% of all public status updates on Twitter — that allows tracking of up to 400 keywords. We collected tweets that contained each of the selected emotion keywords, storing those which do not include a filtered phrase into a MySQL database. We programmed the PHP script to be cyclical in the sense that it streamed individual tweets, but changed emotion keywords every 5 minutes in order to collect the whole range of emotions. Ten days of data collection resulted in a labelled Temporal Emotion Database containing six emotion theories totaling to 21 unique emotion keywords each with at least 1100 documents to base our analysis on. It should be noted that by using WORDNET [29] we could have expanded our initial list of 21 keywords by taking synonyms of each keyword and testing the stream rate for each emotion, selecting the most popular keyword; however in order to fairly test each theory, we opted against this as the selected emotion keywords had been carefully chosen by each theorist.

5. Semantic Emotion Analysis

Having created an emotional Twitter corpus, we analysed this data in order to evaluate the semantic distinctiveness of existing basic emotion sets, developing an iterative latent semantic clustering algorithm to discern the optimal semantically irreducible basic emotion set from all 21 emotions collected. Latent Semantic Clustering (LSC) is a simple modification of the LSA algorithm which we base our DELSAR algorithm on. Given a labelled corpus C with label set K, it calculates, using LSA, the semantic accuracy of each $label \in K$, thus providing an analysis of how distinct the labelling of C and the selection of K is. All analysis was performed on an Intel Core 2 U7700 CPU 2x1.33GHz with 2GB RAM using the GENSIM framework for Python [30] to create LSA spaces. Unless specified, we tested dimensions of the LSA space in increments of 10 and selected the dimensionality that performed optimally for each task, similar to Recchia [31]. For all tasks, we use Log-Entropy normalisation as our Association Function, found to generate optimal results [32] and recommended for LSA [23].

5.1. *DELSAR*

Document-Emotion Latent Semantic Algorithmic Reducer (DELSAR) takes an emotion set and clusters each document's emotion to the emotion of its closest document vector (excluding itself), calculating a clustering accuracy for each emotion. The closest document vector is calculated as the maximum cosine value of the angle between the current document and each other document in the subcorpus. The emotion keyword in each document is removed before the closest document vector is calculated, so we focus purely on the words surrounding the emotion keyword for each document. DELSAR operates in the LSA space created from the subcorpus of all documents matching all emotion keywords in the set being analysed, in which there are ($doc\_limit \times number\_of\_emotions$) documents.

If a document expressing a certain emotion, e, is not clustered with a document of the same emotion, then the words surrounding e is more similar to the words surrounding another emotion. Thus the clustering accuracy of an emotion set corresponds to how distinct that emotion set is; the more semantically accurate an emotion set is, the more distinct the language surrounding each emotion within the set is.

The reduction aspect of DELSAR initially starts with the set of all 21 emotions. After calculating the clustering accuracies for each emotion, it removes the least accurate emotion from the set and iterates until there are n emotions remaining in the initial set, resulting in the optimal semantically distinct basic emotion set. The DELSAR algorithm is described in Algorithm 5.1.

Algorithm 5.1 DELSAR

Require: Final keyword set size *reduceTo*, Corpus **C** and Keyword Set **K**, where
  $\forall document \in$ **C** $\exists document \rightarrow emotion \in$ **K**
  calculate cosine document similarity matrix of LSC(**C**, **K**)
  for each *document* $\in$ **C do**
    delete *emotion* in *document*
    Find closest document vector *nearest* where *nearest* $\neq$ *document*
    if
    nearest(**K**) $==$ *document*(**K**) **then**
      document is a hit
    else
      document is a miss
    end if
  end for each
  for each *emotion* $\in$ **K do**
    calculate accuracy of *emotion* using (total *document* hits where *emotion* in *document*/total *document* where *emotion* in *document*)
  end for each
  if
  length(**K**) $>$ *reduceTo* **then**
    delete least accurate *word* in **K**
    DELSAR(reduced **K**)
  else
    return K
  end if

We performed DELSAR1000 on the corpus and various subcorpora and report the results in Table 2. Note that DELSAR creates an LSA space of all documents within each emotion set; for each basic emotion set an LSA space of $(1000 \times number\_of\_emotions)$ documents is created. Evaluating all sets, our results show the accuracy of clustering each document to its nearest document, whether it is the same or another emotion. Of all the theories analysed, Ekman's set proved to be the most semantically distinct, with a 2.9% increase in accuracy compared to the average of the remaining sets. Russell's categories performed worst, which is surprising seeing as these emotions were taken as the basis for representing the entire emotion spectrum as a whole.

Table 2 DELSAR clustering accuracy of each basic emotion set using a corpus comprised of 1000 documents for each emotion within each set. Standard Deviation of all models is $\sigma = 0.027$.

Dimension	Model							
	Izard 40	Russell 30	Plutchik 30	Ekman 30	Tomkins 30	Oatley 30	All 60	DELSAR 40
accepting			0.583				0.452	0.553
angry	0.390	0.409	0.400	0.429	0.391	0.468	0.248	
anticipating		0.455					0.312	
anxious						0.535	0.272	
ashamed	0.452				0.467		0.366	0.534
contempt	0.550				0.575		0.356	0.574
depressed		0.292					0.193	
disgusted	0.364		0.417	0.484	0.422	0.527	0.227	
excited		0.407					0.227	
guilty	0.426						0.339	
happy						0.411	0.255	
interested	0.561				0.560		0.460	0.603
joyful	0.482		0.518	0.565	0.507		0.397	0.519
miserable		0.413					0.272	
pleased		0.548					0.359	0.506
relaxed		0.383					0.245	
sad			0.388	0.442		0.424	0.259	
scared	0.377		0.456	0.498	0.396		0.249	
sleepy		0.445					0.332	0.591
stressed	0.454	0.376			0.481		0.295	0.502
surprised	0.416		0.491	0.505	0.414		0.295	
MEAN	0.447	0.409	0.464	**0.487**	0.468	0.473	0.306	**0.548**
STDEV	0.068	0.072	0.065	0.049	0.069	0.057	0.072	0.039

We performed DELSAR on the set of all 21 emotions, reducing the set to the eight most semantically distinct dimensions of emotion, these being:

ACCEPTING ASHAMED CONTEMPT INTERESTED JOYFUL PLEASED SLEEPY STRESSED

This set achieved a significant increase in terms of accuracy over Ekman's set of 6.1%; we could say that these emotions best represent the emotion spectrum in its entirety, or in other words, the remaining emotions could be expressed as a combination or a particular degree of intensity of these emotions.

In addition to performing DELSAR1000, we tested four subcorpora of varying document sizes to observe any temporal effects and found negligible temporal variance within our results [16].

5.2. *ELSA*

While DELSAR is highly effective, its analysis is relative to a subcorpus of documents that express all the emotions contained within a particular basic emotion set — whilst it is a good measure of showing how distinct a particular emotion set is overall, it does not allow for each emotion to be mutually independent. This is important to take into consideration as it allows us to compare emotions without the constraint of it being in a set with other emotions — for example an emotion within a set may be considered distinct only because other emotions within the set are not. Emotional Latent Semantic Analysis (ELSA) is a modified version of DELSAR, in which emotions are treated separately from one another. ELSA takes the set of all 21 emotions and, *for each emotion*, creates an LSA space using documents matching only that particular emotion, in which there are (*doc_limit*) documents. For each ELSA space, the cosine value for the closest document vector to each document is determined, and an average of these is calculated. The higher this average value is for a specific emotion, the more similar the documents are for that emotion, in other words, the emotion cluster is tightly packed. Lower values mean less similar words being used in the expression of the same emotion — the emotion cluster is more dispersed — signifying a decrease in distinctiveness. The difference between ELSA and DELSAR, is that the latter evaluates whether a particular emotion *set* is representative of the entire emotion spectrum, as opposed to seeing which *emotions* are distinct.

Evaluating each basic emotion set according to ELSA is simply a matter of averaging the corresponding values of the constituent emotions, and discerning the most semantically distinct emotions requires selecting the emotions with maximum average values. The ELSA algorithm is described in Algorithm 5.2.

Algorithm 5.2 ELSA

Require: Corpus **C** and Keyword Set **K**, where ∀*document* ∈ **C** ∃*document* →
  emotion ∈ **K**
  for each *emotion* ∈ **K do**
    for each *document* ∈ **C do**
      if
      document(**K**) == *emotion* **then**
        delete *emotion* in *document*
        calculate cosine document similarity matrix of LSA(*document*, **C**)
        Find closest document vector *nearest* where *nearest* ≠ *document*
      end if
    end for each
    return average(*nearest*)
  end for each

We performed ELSA in a similar fashion to DELSAR — testing the same copora — and report the results in Table 3. Out of all the basic emotion sets analysed, Tomkin's set proved to contain the most semantically concentrated emotions, although it must be pointed out that Tomkin's set is identical to Ekman's set without the emotion *sad* and four other emotions added; by swapping *disgusted* for *contempt*, Ekman's set would have been optimal at 0.747.

Table 3 ELSA average cosine values using dimensions 10, 20, 30, 40, 50, 60, 70, 80, 90 and 100. Each emotion uses a corpus of 1000 documents. Standard Deviation of all models is $\sigma = 0.010$.

	Model							
	Izard	Russell	Plutchik	Ekman	Tomkins	Oatley	All	ELSA
accepting			0.781				0.781	0.781
angry	0.727	0.727	0.727	0.727	0.727	0.727	0.727	
anticipating			0.717				0.717	
anxious						0.744	0.744	0.744
ashamed	0.743				0.743		0.743	0.743
contempt	0.838				0.838		0.838	0.838
depressed		0.695					0.695	
disgusted	0.708		0.708	0.708	0.708	0.708	0.708	
excited		0.708					0.708	
guilty	0.713						0.713	
happy						0.694	0.694	
interested	0.724				0.724		0.724	
joyful	0.761		0.761	0.761	0.761		0.761	0.761
miserable		0.744					0.744	0.744
pleased		0.742					0.742	0.742
relaxed		0.707					0.707	
sad			0.713	0.713		0.713	0.713	
scared	0.719		0.719	0.719	0.719		0.719	
sleepy		0.704					0.704	
stressed	0.736	0.736			0.736		0.736	0.736
surprised	0.723		0.723	0.723	0.723		0.723	
MEAN	0.739	0.720	0.731	0.725	**0.742**	0.717	0.731	**0.761**
STDEV	0.038	0.019	0.026	0.019	0.039	0.019	0.033	0.034

We obtained a slightly different optimal set consisting of the eight most semantically distinct emotions compared to DELSAR, taking away *interested* and *sleepy* and adding *anxious* and *miserable*:

ACCEPTING ANXIOUS ASHAMED CONTEMPT JOYFUL MISERABLE PLEASED STRESSED

This set achieved a 1.9% increase in accuracy compared to Tomkin's set. We could say that this basic emotion set contains those emotions which are the most *atomic* in the sense that the words surrounding these emotion keywords are semantically concentrated; people using these emotions are more likely to be actually referring to these emotions due to the similarity of language across all documents. Take *happy* as an example, which is the least atomic emotion: being the least semantically concentrated means that the language that people use when using the word *happy* varies the most, either due to describing a great variety of things, being used in a great variety of contexts, or varying perceptions of what the emotion *happy* actually means.

6. Conclusion

A vast majority of computer scientists tend to use Ekman's basic emotion set for emotion categorisation, and it appears that, semantically, it is the most distinct set, with a 2.9% increase in accuracy compared to the average of the remaining sets. Using an iterative algorithm based on LSC, we have discerned a set of eight (rather than Ekman's six) basic emotion keywords that have been calculated to be the most semantically distinct. This set performed better in all semantic tests than all of the basic emotion models analysed, with a 6.1% increase in accuracy over Ekman's basic emotion set, providing evidence that by carefully selecting emotion keywords, more of the emotion spectrum can be accounted for. It must be noted, however, that the lack of varience of surrounding words of identified basic emotions may just depict a stricter consensus on the *definition* of the word, unrelated to any emotional phenomenological hierarchy.

Emotions must be seen as relative to a specific domain; it has been recently shown that facial expressions of emotion are not culturally universal [33]. Basic emotions are ultimately not universal and are correlated with underlying thematic concerns within the corpus under analysis.

Ranking emotions and basic emotion sets using our algorithms according to a metric of semantic distinctiveness allows us to compare the similarity of compound emotions, analyse the composite properties of emotions and highlight how specific emotions interact with each other, with applications ranging from clinical assessments to emotion engineering.

Emotion may contribute to evolution on a much grander scale than previously thought. Indeed, Izard [13] suggests that the main component in evolution could be Emotion Schemas, that is, evolution of actions through imitative learning of specific emotions. Memetic theory states that the ability to imitate is the only requirement for language to occur in evolution, and it has been shown in several studies that syntax and semantics emerge spontaneously (for a discussion, see Blackmore [34]). Thus, by analysing language we should be able to reverse-engineer the imitative mechanisms of humans. Mapping such processes could shed light on an updated and, combined with genetic algorithms, a more complete model of human evolution.

Acknowledgements

We would like to thank Paul Rauwolf and Yifei Wang for their helpful comments.

References

1. J. Watson, *Behaviourism* (University of Chicago Press, 1930).
2. C. E. Izard, *The Face of Emotion* (Appleton-Century-Crofts, New York, 1971).

3. R. Plutchik, *A general psychoevolutionary theory of emotion*, in *Emotion: Theory, research, and experience: Vol. 1. Theories of emotion*, eds. R. Plutchik and H. Kellerman (Academic press, New York, 1980), New York, pp. 3–33.

4. J. Panksepp, *Behavioral and Brain Sciences* **5**, 407 (1982).

5. J. Gray and N. McNaughton, *Nebraska Symposium on Motivation* **43**, 61 (1996).

6. A. Ortony and T. J. Turner, *Psychological Review* **97**, 315 (1990).

7. L. F. Barrett, *Personality and Social Psychology Review* **10**, 20(February 2006).

8. K. A. Lindquist and L. F. Barrett, *Psychological Science* **19**, 898 (2008).

9. Aristotle, *Rhetoric* 350BC.

10. P. Ekman, *Emotions Revealed: Understanding Faces and Feelings* (Phoenix, 2004).

11. J. Panksepp, *The Caldron of Consciousness: Motivation, affect and self-organization - An anthology (Advances in Consciousness Research)* 2000.

12. A. Damasio, *The Feeling of What Happens: Body Emotion and the Making of Consciousness.* (Vintage, 1999).

13. C. E. Izard, *Annual review of psychology* **60**, 1 (2009).

14. D. Dennett, Quining qualia, in *Consciousness in Contemporary Science*, eds. A. J. Marcel and E. Bisiach (Oxford University Press, Oxford, 1988) pp. 42–77.

15. P. Ekman, W. V. Friesen and P. Ellsworth, *Emotion in the Human Face* (Oxford University Press, 1972).

16. E. Y. Bann, *Discovering Basic Emotion Sets via Semantic Clustering on a Twitter Corpus*, tech. rep., University of Bath (2012), www.aeir.co.uk/pub.

17. S. D. Rosenberg and G. J. Tucker, *Archives of General Psychiatry* **38**, 1331 (1979).

18. J. W. Pennebaker and L. A. King, *Journal of personality and social psychology* **77**, 1296(December 1999).

19. M. L. Newman, J. W. Pennebaker, D. S. Berry and J. M. Richards, *Personality and Social Psychology Bulletin* **29**, 665 (2003).

20. A. Bilovich and J. J. Bryson, Detecting the evolution of semantics and individual beliefs through statistical analysis of language use, in *Naturally-Inspired Artificial Intelligence - Papers from the AAAI Fall Symposium*, 2008.

21. F. R. Gill, A.J., Level of representation and semantic distance: Rating author personality from texts, in *Proc. Euro Cogsci*, 2007.

22. A. J. Gill, D. Gergle, R. M. French and J. Oberlander, Emotion rating from short blog texts, in *Proceedings of the SIGCHI Conference on Human Factors in Computing Systems*, CHI '08 (ACM, New York, NY, USA, 2008).

23. T. K. Landauer and S. T. Dumais, *Psychological Review* (1997).

24. C. Strapparava and R. Mihalcea, Learning to identify emotions in text, in *Proceedings of the ACM Conference on Applied Computing*, March 2008.

25. A. J. Gill, D. Gergle, R. M. French and J. Oberlander, Emotion rating from short blog texts, in *Proceedings of the 26th Annual SIGCHI Conference on Human Factors in Computing Systems*, 2008.

26. J. A. Russell, *Journal of Personality and Social Psychology* **39**, 1161 (1980).
27. J. Bollen, A. Pepe and H. Mao, *CoRR* **abs/0911.1583** (2009), informal publication.
28. S. M. Kim, A. Valitutti and R. A. Calvo, Evaluation of unsupervised emotion models to textual affect recognition, in *Proceedings of the NAACL HLT 2010 Workshop on Computational Approaches to Analysis and Generation of Emotion in Text*, (Association for Computational Linguistics, Los Angeles, CA, June 2010).
29. C. Fellbaum (ed.), *WordNet: An Electronic Lexical Database (Language, Speech, and Communication)* (The MIT Press, May 1998).
30. R. Řehůřek and P. Sojka, Software Framework for Topic Modelling with Large Corpora, in *Proceedings of the LREC 2010 Workshop on New Challenges for NLP Frameworks*, (ELRA, Valletta, Malta, May 2010).
31. G. Recchia and M. N. Jones, *Behavior Research Methods* **41**, p. 647 (2009).
32. P. Nakov, A. Popova and P. Mateev, Weight functions impact on lsa performance, in *EuroConference RANLP'2001 (Recent Advances in NLP*, 2001.
33. R. Jack, O. Garrod, H. Yu, R. Caldara and P. Schyns, Facial expressions of emotion are not culturally universal, in *Proceedings of the National Academy of Sciences of the United States of America*, May 2012. in press.
34. S. Blackmore, *Journal of Consciousness Studies* **10**, 19 (2003).

A NEURO-COMPUTATIONAL STUDY OF LAUGHTER

M. F. ALONSO, P. LOSTE, J. NAVARRO, R. DEL MORAL

Bioinformation and Systems Biology Group. Instituto Aragonés de Ciencias de la Salud.
Zaragoza, Spain

R. LAHOZ-BELTRA

Department of Applied Mathematics, Faculty of Biological Sciences, Complutense
University of Madrid
Madrid, Spain

P. C MARIJUÁN

Bioinformation and Systems Biology Group. Instituto Aragonés de Ciencias de la Salud.
Zaragoza, Spain

A new core hypothesis on laughter is presented. It has been built by putting together ideas from several disciplines: neurodynamics, evolutionary neurobiology, paleoanthropology, social networks, and communication studies. The hypothesis contributes to ascertain the evolutionary origins of human laughter in connection with its cognitive and emotional signalling functions within social groups. The new behavioural and neurodynamic tenets introduced about this unusual sound feature of our species justify the ubiquitous presence it has in social interactions and along the life cycle of the individual. Laughter, far from being a curious evolutionary relic or a rather trivial innate behaviour, should be considered as a highly efficient tool for inter-individual problem solving and for maintenance of social bonds. In spite of a number of recent studies, links between different classes of laughter and emotional conditions of individuals have not been clearly established yet. There is also a lack of research about the presence of emotional patterns in spontaneous laughing sounds, which could provide keys to a subsequent classification. There are no studies in psychiatry attempting the use of laughter as a diagnostic tool either. In order to advance laughter research in these aspects, the present study has extracted nine temporal and acoustic variables from laughter episodes, looking for establishing better pattern-classification methods. These variables are number of bouts, voiced percentage, bout duration, mean energy, mean entropy, standard deviation of fundamental frequency, jitter, shimmer and harmonic to noise ratio. A library of records of laughter episodes was created and the values of the previous variables were statistically analyzed. In spite of the variability of acoustical parameters, there were significant differences which might be associated to the different laugh types. In future works, these statistical outcomes will be contrasted with the evaluations by human subjects.

1. Introduction: The evolutionary origins of laughter and the social brain hypothesis

The revival of laughter research during last two decades [1] has been very fertile concerning specialized achievements in the neuroimaging, neurophysiological, sound analysis, physiological (respiratory & phonatory), ethological, evolutionary, social and health aspects related to laughter. However, the conceptual counterpart of putting together the most relevant strands of thought in order to gain more advanced synthetic views or even to establish a new core hypothesis has not been developed sufficiently. The preliminary idea of this paper is to establish a coherent link between the evolutionary roots of laughter and the origins of language, aligned with the "social brain" hypothesis[2,3], and further to connect these views with the "sentic forms" hypothesis.

The social brain hypothesis posits that, in primate societies, selection has favoured larger brains and more complex cognitive capabilities as a mean to cope with the challenges of social life [3,4]. Social networks in primates seem to be very different from those found in most other mammals: they are cognitive, memory-loaded, based on bonded relationships of a kind found only in pairbonds of other taxa[5]. Frequent pair-wise grooming in between individuals, however, imposes a strict time limitation regarding group size: depending on diet, 20% of time is the upper ecological limit that grooming can reach. This factor necessarily restricts the size of grooming networks and, thus, of natural groups in primate societies (composed, at most, of a few dozen individuals).

So, how could human societies have organized their "grooming" within increasingly larger natural groups, of around 100 or 150 individuals? As Dunbar[3,6] has argued, human language was the evolutionary solution. "Languaging" was co-opted as a *virtual system* for social *grooming-massaging*, plus other specifically human adaptations for group cohesion: laughter, crying, gaze-facial expressions, music, dance... It is by following this line of thought, that the enigmatic presence of laughter along the human life cycle may be further clarified.

Laughter quite often breaks in amidst the talking/listening exercise. Having evolutionarily preceded language, laughter has continued to fulfill very especial tasks regarding the communicational grooming of human groups. What has been called "antiphonal laughter" (the chorus of laughing people[7,8]) may be seen as an *effective extension* of the talking massage effects in bigger groups, where the mere size precludes active participation of most individuals in the talk; the laughing together that ensues, brings the augmented neuromolecular grooming-effects of laughter available to everybody in the group irrespective of the conversation share.

Laughter is regularly situated at the very end of verbal utterances; it punctuates sentences as a sort of emotional valuation or as an enigmatic social "call", even in deaf people using the hand-sign language[9]. In this sense, laughter production, far from interfering with language or competing as a "low level" process with the higher cognitive functions for access to the fonatory apparatus, becomes itself a *cognitive solution*, marking the occurrence of humorous incongruences as positively finalized items within the ongoing talking/listening exchange[10-12]. This is the core of the "neurocomputational hypothesis" we are going to develop in the rest of this paper.

Far from implying a reductionism view of laughter, an ampler framework is proposed herein, where laughter becomes an *abstract* information processing tool that can influence a number of emotional, cognitive, and social contexts.

2. Sound structures of laughter

Far from being a stereotyped signal, laughter becomes one of the most variable acoustic expressions of humans, comparable to language except for the combinatory richness of the latter. Typical laughter is composed of separate elements or "calls" or "syllables", *plosives*, over which a vibrato of some fundamental frequency Fo is superimposed[13]. A typical laughter episode may last around one second (or slightly less) and will contain around 5 plosives (more in general, in between 2 and 8). An important distinction to make is between "vocalized" and "unvocalized" laughter; even though the former induces significantly more emotional responses in listeners, the latter appears consistently in many laughter records, comprising a large variety of sounds (snort-like, grunt-like, giggles, chuckles, etc.).

In a landmark experimental study, Bachorowski *et al.*[14] found that there are around 4.4 calls or plosives within each laughter bout, a single plosive having a duration of 0.11 s and a separating interval of 0.12 s (for voiced laughter). Call or plosive production is denser towards the beginning of laugh bouts, and inter-plosive durations gradually increase over the course of bouts. The average value of the fundamental frequency Fo for male laughs is 272 Hz (sd = 148) while for females is considerably higher and more variable 405 Hz (sd 193); only for voiced laughs, the respective values are 282 and 421 Hz. Usually Fo is much higher in laughter than in speech, thus, extremes of male Fo were found to be as high as 898 and as low as 43 Hz, while female extremes were in between 2083 and 70 Hz. The excursions of Fo along the bout trajectory represent an additional factor of variability, showing contours such as "flat", "rising", "falling", "arched", sinusoidal", etc.

All of the previous elements could form part of the inbuilt cues to *laugher identity*, which have been proposed to play an important role in listener emotional responses[15]. In particular, the pitch or *tone* curve described by *Fo*, together with the distribution of plosives, would show consistent differences between laugh forms associated with emotional states of positive and negative valence[16]. The main trend is that the energy and duration becomes higher for "positive" than for "negative" laugh, and vice versa for the relative presence of unvoiced frames, more frequent in ironic and hostile laughs than in joyful ones. Notwithstanding that, there is not much consensus established yet –neither significant hypothesis to put to test– on how the interrelationship between plosives, tones, melodies and other variables of laughter may be systematically involved in encoding and distinguishing the underlying emotional states[11,12,17].

2.1. *Auditory patterns processing*

The acoustical patterns received are processed in several centres depending on the type of information. There are two main theories in the path of pitch recognition: the first one begins with the tonotopic organization of the cochlea that continues in the basilar membrane at different levels depending on the frequency perceived, hence activating neurons with different characteristics frequencies; and the path continues to the cortex. The second theory assumes that the pitch of a stimulus has a excitation pattern associated with the pure tone of the stimulus[18]. However, animal studies with electrophysiological measures suggest that the primary auditory cortex does not contain the pitch representation[19,20]. Further, a cortical zone of pitch processing was observed in human with fMRI in the anterolateral region of the auditory cortex[21,22]. Moreover, a primate study shows the existence of neurons that respond to pure tones and to complex sounds without the fundamental harmonic in the same anterolateral region of the auditory cortex area, thus discarding the pitch perception[23].

According to Szameitat´s observations the acoustical processing of tickling and emotional laughs are different[24,25]. The tickling has a higher information transfer rate and higher acoustical complexity than emotional laughter. The perception of emotional laughs generates more activation in the rostral prefrontal cortex than the perception of tickling laugh. In tasks of explicit tone processing of emotional laugh, the hemodynamic increased in the right posterior superior temporal sulcus (STS), bilateral inferior fronto-lateral gyrus (IFG) and orbito-frontal cortex (OFC), and the bilateral posterior rostral medial frontal cortex (prMFC). These areas are related with auditory visual processing (STG), working memory and decision making (IFG), emotional evaluation (OFC) and explicit speech melody evaluation (prMFC)[26]. Then, several networks are activated depending of the type of humour to which the subject is exposed to: semantic juxtapositions (incongruity) use a bilateral temporal lobe network, whereas phonological juxtapositions (puns) uses a left hemisphere network centered on speech production regions.

There is a separate and distinct network for the affective components of humour, with mainly high activation in medial ventral prefrontal cortex and in bilateral cerebellum[25]. The affective prosody processing has been linked to the mirror system; this neuronal system is related with action understanding and recognition tasks, with auditory, visual and motor stimulus. The mirror neuron system plays a crucial role in high level meta-function like language, theory of mind and empathy[27]. Through animal studies it has been observed that the areas involved in this system are: the ventral pre-motor area of the IFG, parietal frontal in the rostral cortical convexity of the inferior parietal lobule (IPL), and STS[28]. The cortical response to musical stimuli suggests that the IFG, mirror auditory system generates a multimodal representation from auditory stimulus[29].

2.2. *Emotional content in our communication: Sentic forms*

The recognition of emotion in speech prosodic patterns is not equal to the acoustical patterns on laughs, but since laughter predates speech both ontogenetic and phylogenetic, the relation is evident. In fact, there is an acoustical correlation of emotional dimensions between speech and laughter, e.g. arousal increases, laughter becomes faster, with higher pitch and lower shimmer and jitter[25]. According to Banse et al.[30], when vocal expressions are analyzed, we find that every emotion has an acoustical profile: anger has higher F_0 mean, energy mean, variability of F_0 across the utterance encoded and the rate of articulation usually increases; in fear high arousal levels are observed with increases of F_0 mean, range and high-frequency energy; joy has increases of F_0, F_0 range, F_0 variability and mean energy; and in disgust the results are inconsistent[30].

In our neurocomputational hypothesis, laughter becomes a quasi-universal information processing "finalizer". We laugh "abstractly": when a significant neurodynamic gradient vanishes swiftly, i.e., when a relatively important problem of whatever type has been suddenly channelled in a positive way, and has vanished as such problem. Like in the slow tension growth and fast release of physical massage, the paradoxical, or tense, or contradictory situation suddenly becomes a well-known case of pleasurable, primary, childish, stumbling, babbling, or retarded-foreigner nature. Problem solved! The "idle" excitation still circulating in the regular problem-solving of cortical and limbic structures is redirected towards the fonatory apparatus where it produces an unmistakable signature. It is the "call" of the species, a social signal of wellness after successful problem solving, after effective mental massage. The sound form of laughter would bear a trace on the kind of neurodynamic gradient that originated it[10,12].

The *sentic forms* hypothesis, framed by M. Clynes in the 70's[31], could help in the exploration of new directions for such open questions. If laughter contains inner "melodies" or pitch patterns of emotional character, how could they be structured? Following the sentic paradigm developed around tactile emotional communication by means of exchange of pressure gradients, there appears a set of universal dynamic forms that faithfully express the emotional interactions of the subjects (see Figure 1)[31,32].

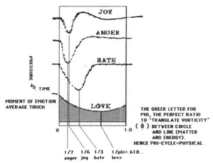

Fig. 1. Dan Winter's mathematical interpretation on various Sentic Forms of M. Clynes (corresponding to four basic emotions).

This approach is in congruence with recent application of theories of embodied or grounded cognition to the recognition and the interpretation of facial expression of emotion. For instance a model of the processing of smile, the most complex facial expression, is presented as a means to illustrate how to advance the application of theories of embodied cognition in the study of facial expression of emotion[33]. In fact, our approach to sentic forms in the sounds of laughter dovetails with the corresponding forms of facial expressions, as should be expected.

3. Experimental Methodology

Checking the hypothesis of sentic forms requires studying a great number of laughs (laughter library). One of the main difficulties to obtain this laughter library is to create an appropriate environment where spontaneous social interactions involving laughter could be recorded; but it is not easy to feel comfortable and laugh naturally when a recorder is capturing all your sounds and expressions. As a first approach, our laugh library has been compiled with internet laugh records and with our own recordings.

3.1. *Compilation of a laughter library*

The compilation of laugh videos, uploaded by subjects themselves, was obtained through the internet. Besides most of these videos provide the circumstance of the triggering event and according to the emotional situation, the laughs may be classified easily.

Several recording sessions were performed with a specific protocol involving numerous kinds of stimuli to generate the different emotions studied. Records were made with a Digital voice recorder, Olympus VN-711PC and Olympus VN-712PC. Directly we registered laughs of more than one hundred individuals, included men and women between the age of 20 and 65, whose laughs were registered individually by a tape recorder. All the individuals were Spanish and none of them suffered any mental illness, so they were able to understand the entire humour sketches. The main emotions induced by these videos were: happiness, hostility, surprise and excitation.

Spontaneous laughter was recorded from each participant and was transformed in a wav archive encoded in 16-Bit PCM format, sampled at 22050 Hz. We separated every laugh event by hearing the recordings and visualizing the waveform using Adobe Audition, a sound analysis program. Through this software we distinguished every laughter episode, that is, an entire laughter utterance, and we stored them separately. Whether a laughter episode was suitable or not, depended on our decision. The validity of the audio segment was conditioned mainly on its clarity; overlapped speech-laugh or laugh-laugh segments were dismissed. All laugh archives had well defined boundaries, were coming from only one individual, and did not have any interfering noise like humming or throat cleaning among others. Otherwise the laughs were discarded. This task is difficult to achieve with automatic laughter detectors, including machine learning methods and support vector machines, although manually it becomes a slow process, it is reliable enough.

On the other hand, we worked with a previously laughter database with isolated laughter bouts, and based on these records we focused on the development of a robust burst-laugh detector. In laughter analysis and characterization, the wide range of acoustic shapes requires segmentation in time domain. To classify human laughter, the hierarchical decomposition method is useful to extract information from rhythm; further features can be computed only on specific segments. According to Trouvain[34] and Dupont[35] each laughter bout consists of a number of elements, called "plosives", characterized by energy peaks separated by silences, repeated in series every 210 ms approx. At this temporal level, bouts can be seen as alternating maxima and minima in waveform amplitude envelope. We don't consider syllables with voiced-unvoiced concatenation as a valid decomposition, such the stereotypical "hahaha", because there are several types of unvoiced laughs with noisy and irregular sounds.

3.2. *Data processing*

The plosive automatic detector was implemented in Matlab. For this purpose we created a program that reads every sound archive and computes the mean energy values of concatenated signal frames, with a frame length of 21 ms and a shift of 5 ms. Due to the wide frequency distribution of unvoiced laughs (snort, grunt, etc), a spectral-domain method isn't adequate. Thereafter, the program finds the lowest energy points using a valley detection algorithm. To confirm if these points are accurately plosive limits, there should be at least 10% of maximum energy between two valleys and duration of 100 ms to discard spurious vocalizations. The number of plosives in an episode varied significantly, for example, one bout can contain in between one and eight plosives. However, to perform straightforward comparisons between laughter types, it is better to build vectors of fixed length. The features extracted in our study for every plosive were:

- *Total time duration in seconds.* Breathing is the primal driving force of laughter and marks a very definite pattern different from other expressions. The evolution of burst durations gives an overview of the temporal structure of a laughter episode.
- *Fundamental frequency mean.* It can be defined as the inverse of the smallest period of the vocal fold signal, in the interval being analyzed (every 10 ms). The algorithm is based on the normalized cross-correlation function. However, since pitch has a very short-time variation, we compute its mean. We consider the fundamental frequency important because vocal articulators are less accurate and slower in nonverbal expressions; therefore vocal chords have a key role in laughter characterization.
- *Standard deviation of the aforementioned fundamental frequency.* It was computed over successive short time frames of the signal, in order to determine range and variability of an instantaneous measurement.
- *First three formants.* A formant is defined as a spectral peak of the spectrum of the voice that gives information about the acoustic resonances of the vocal tract position at every plosive. To estimate formants, we used a method based on averaging different linear predictive code spectrograms, and detected the first three maximum values. Vocal

articulators act as filters of a resonant tube, changing the fundamental frequency. When communicating different emotions or intentions these articulators are a key mechanism for changing emotional information (in general, emotional expression is dominated by effects created in the glottis and laryngeal muscles).

- *Average power or energy per sample.* The amplitude of sound affects our perception of fundamental frequency, what is called pitch. The pitch is an interpretation of the listener, and one of the parameters that influence this subjective interpretation is the intensity or sound energy.

- *Shannon's entropy.* This feature measures the variable information contained in the patterns of a message, as opposed to the portion of the message that is determined.

- *Jitter*, a cycle-to-cycle measure of the variations of fundamental frequency, as a percentage. It is known that emotions are expressed in many cases as unintentional acts, such as quiver or tremble. Jitter is defined as vocal chords disturbances that occur during the glottal phonation of the vowel and affect the pitch.

- *Shimmer*, defined as the average absolute difference between the amplitudes of consecutive periods, expressed as a percentage. It is equivalent to jitter in signal amplitude, that is, rapid variations in loudness.

- *Harmonic to Noise Ratio (HNR)*, described as the energy ratio of fundamental frequency harmonics to noise components. HNR appears to be a more sensitive index of vocal function than other features.

- *Percentage of voiced signal over unvoiced signal*, that is, the time that vocal cords vibration spends over a plosive. This measure shows how much value has the harmonic expression from the vocal cords, as opposed to sound generated in other parts of the vocal tract.

The outcome of this characterization is a data matrix consisting of all plosives sorted by laugh archives in rows, with 12 columns per plosive. In order to fit statistical pattern recognition for every laugh event, we set out a matrix with one column per archive, inserting all features in one row. This arrangement yields a huge amount of data to process and classify. Obviously, we had to compress data restricting plosives to five, and discarding proved weak features to classification such as jitter, shimmer, HNR and average power.

4. Preliminary Results

A discriminant analysis was performed. Table 1 shows the results of the discriminant analysis of data classified into three type groups: happiness (0), hostility (1) and surprise (2). The table shows the results correctly classified for 129 laugther samples of the different laugh types, indicating that 85.14% of the samples are correctly identified in happiness, 85.42% in hostility and 85.71% in surprise. The first two discriminant functions are statistically significant (P < 0.001 and P < 0.01 respectively) accounting both for a cumulative percentage 100% of the variance. In consequence, according to three categories of laugh, there are well-defined groups of each laugh type, corresponding with different emotional states, and they can be successfully classified by analyzing each laughter. A high value of the canonical correlation in the first (0.74) and second (0.61) discriminant functions indicate a high relationship between the group membership and discriminant function values. This fact is also manifested in the first (eigenvalue equal to 1.21) and second (eigenvalue equal to 0.26) discriminant function. Likewise each discriminant function stands for a proportion of 67% and 33.32% of the total variance (Wilk's Lambda) not explained by differences between groups.

Table 1. The table shows the results of discriminant analysis classified into three laugh type groups: happiness (0), hostility (1), and surprise (2).

		LAUGH TYPE		
	N	Type 0	Type 1	Type 2
Type 0	74	63 (85,14%)	8 (10,81%)	3 (4,05%)
Type 1	48	6 (12,50%)	41 (85,42%)	1 (2,08%)
Type 2	7	1 (14,29%)	0 (0,00%)	6 (85,71%)

In the figure below (Figure 2), we can see that there are well-defined groups and hence an emotional state corresponds to a type of laughter

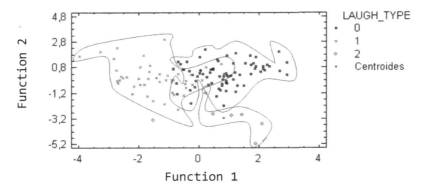

Discriminant Function Analysis

Fig. 2. The graph represents the discriminant function analysis. It shows the results correctly classified in the different laugh types, indicating that 85.14% of the samples are correctly identified in happiness, 85.42% in hostility and 85.71% in surprise.

5. Discussion and Comments

This is an ongoing work and the complete results will be presented in an incoming paper (del Moral et al., in preparation). A library of natural laughs has been compiled herein, so that an artificial neural network should be able to distinguish and categorize the concerned sound structures, gauging the soundness and conceptual possibilities of the sentic forms hypothesis. We were able to find patterns in acoustical features between the different laugh-types according to the expressed emotion. As it may be expected, the differences would be in the features corresponding to the vibration cycle in the pharynx, and they translate in different acoustical patterns that approximately coincide with the sentic forms proposed by Manfred Clynes[31,36].

Of course, that the sentic hypothesis and the whole neurodynamic scheme herein presented become acceptable as a heuristic device is a highly debatable matter (even more in its connection with the social brain hypothesis). But the commonality between these views is remarkable: the global/local entropic variable comprising the evolution of brute excitation, which is shared by the different motor expression capabilities and easily recognizable by all sensory modalities; the centrality of the gradients and the different dynamic form corresponding to each recognizable emotion. Clynes himself wrote about laughter as "another sentic form"[31,36], or as a *composite* of sentic forms --as we would mean here. Beyond the particulars of laughter, let us remark that a number of illustrious voices in contemporary neuroscience could be enlisted in support of the need of new synthetic theories about human information processing, perhaps not too distant from the argumentative lines herein discussed[10,37-40].

In the extent to which this sentic scheme is acceptable, or at least permissible as a heuristic approach, it can throw light on why humans have evolutionarily augmented the innate behavior of laughter (as well as crying and other group emotional adaptations). Laughter is spontaneously produced to minimize occurring problems in an automatic-unconscious way that mobilizes powerful neurodynamic and neuromolecular resources without any extra computational burden on the ongoing conscious processes of the individual. In the complex social world that the enlarged human brain confronts, with multitude of perceptual, sensorimotor, and relational problems, and above all with those derived from the conceptual-symbolic world of language in the making and breaking of social bonds, informational problems dramatically accumulate in very short time spans. Thus, it makes a lot of evolutionary sense counting with these extra-ordinary minimization resources: the information processing power of a hearty laugh (or of bursting out into tears!).

Laughter is one of the most complex behaviours exhibited by humans. It integrates the innate and the cultural, the emotional and the cognitive, the individual and the social. Any unifying hypothesis is forced to contain an unwieldy heterogeneity of elements, even in order to attempt a very rudimentary "closure". Some of these elements may locate in well-trodden disciplinary paths and are relatively easy to discuss, while others neatly belong to the theoretical-speculative (at the time being) and become relatively disciplinary-independent. All of them, but particularly the latter, are in need of more meticulous experimental approaches.

References

1. R. R. Provine, *Laughter: A scientific investigation* (Faber & Faber, London, 2000).
2. J. M. Allman, *Evolving Brains* (Scientific American Library, New York, 1999).
3. R. Dunbar, *The Human Story: A New History of Mankind's Evolution* (Faber & Faber Ltd., London, 2004).
4. J. B. Silk, *Science* **317**, 1347 (2007).
5. R. Dunbar and S. Shultz, *Science* **317**, 1344 (2007).
6. R. Dunbar, *Evolutionary Anthropology* **5**, 178 (1998).
7. M. Smoski and J. A. Bachorowski, *Cognition & Emotion* **17**, 327 (2003).
8. M. Smoski, The Development of Antiphonal Laugther Between Friends and Strangers. PhD dissertation Vanderbilt University. Psychology, (Nashville, August, 2004).
9. R. R. Provine and K. Emmorey, *Stud. Deaf. Educ.* **11**, 403 (2006).
10. K. P. Collins and P. C. Marijuán, *El Cerebro Dual: Un Acercamiento Interdisciplinar a la Naturaleza del Conocimiento Humano y Biológico* (Editorial Hacer, Barcelona, 1997).
11. J. A. Bea and P. C. Marijuán, *Entropy* **5**, 205 (2003)
12. P. C. Marijuán and J. Navarro, "The Bonds of Laughter: A Multidisciplinary Inquiry into the Information Processes of Human Laughter". (2010). Cite as: arXiv:1010.5602v1 [q-bio.NC]
13. H. Rothgänger, Comparison on standarized fundamental frequency of crying and lalling, 6th. International Workshop on Infant Cry Research. (Bowness. 1997).
14. J. A. Bachorowski, M. Smoski and M. J. Owren, *J Acoust. Soc. Am.* **110**, 1581 (2001).
15. J. A. Bachorowski and M. J. Owren, *Psychol Sci.* **12**, 252 (2001).
16. L. Devillers and L. Vidrascu. Positive and Negative emotional states behind the laughs in spontaneous spoken dialogs. Interdisciplinary Workshop on The Phonetics of Laughter. Saarbrücken, (Germany, 4-5 August 2007).

17. J. A. Bachorowski. and M. J. Owren. *Sounds of Emotion 1 Vocal Expressions of Emotion*. (Guilford, New York, 2008).
18. P. R. Hill, D. E. Hartley, B. R. Glasberg, B. C. Moore and D. R. Moore, *J Speech Lang Hear Res.* **47,** 1022 (2004)
19. Y. I. Fishman, D. H. Reser, J. C. Arezzo and M. Steinschneider, *Brain Res.* **786**, 18 (1998).
20. D. W. Schwarz and R. W. Tomlinson, *J Neurophysiol.* **64**, 282 (1990).
21. R. D. Patterson, S. Uppenkamp, I. S. Johnsrude and T. D. Griffiths, *Neuron* **36,** 767 (2002).
22. H. Penagos, J. R. Melcher and A. J. Oxenham, *J Neurosci.* **24**, 6810 (2004).
23. D. Bendor and X. Wang *Nature* **436**, 1161 (2005).
24. D. P. Szameitat, B. Kreifelts, K. Alter, A. J. Szameitat, A. Sterr, W. Grodd and D. Wildgruber D, *Neuroimage* **53**, 1264 (2010).
25. D. P. Szameitat, C. J. Darwin, D. Wildgruber, K. Alter and A. J. Szameitat, *Cogn Emot.* **25**, 599 (2011).
26. V. Goel and R. J. Dolan, *Nat Neurosci.* **4**, 237 (2001).
27. V. Ramachandra, *Percept Mot Skills* **108**, 30 (2009).
28. G. Di Pellegrino, L. Fadiga, L. Fogassi, V. Gallese, G. Rizzolatti, *Exp Brain Res* **91,** 176 (1992).
29. A. D'Ausilio, *J Neurosci.* **27**, 5847 (2007)
30. R. Banse, K. R. Scherer, *J Pers Soc Psychol.* **70,** 614 (1996).
31. M. Clynes, *Somatics* **2**, 3 (1979).
32. D. Winte, How to touch: How the Geometry of Pressure in Your Touch-Predicts the Harmonics of Emotion in Music-& Love. (1999) http://soulinvitation.com/touch/touch.html
33. P. M. Niedenthal, M. Mermillod, M. Maringer and U. Hess, *Behavioral and Brain Sciences* **33**, 417 (2010).
34. J. Trouvain. "Segmenting Phonetic Units in Laughter", proceedings of the 15th International congress of phonetic sciences, (Barcelona, 2003).
35. S. Dupont, T. Dubuisson , J. A. Mills, A. Moinet, X. Siebert X, D. Tardieu and J. Urbain, *QPSR of the numediart research program*, **2**, nbr. 2 (2009).
36. M. Clynes, Generalized Emotion: How it May be Produced, and Sentic Cycle Therapy. In: Manfred Clynes & Jaak Panksepp. Editor. Emotions and Psychopathology (1988).
37. G. M. Edelman and G. Tononi, *A Universe of Consciousness* (Basic Books, NY, 2000).
38. G. Buzsáki, *Rhythms of the Brain* (Oxford University Press, New York, 2006).
39. S. Dehaene, *Reading in the Brain.* (Penguin, New York, 2009).
40. O. Sporns, *Comput. Neurosci.*, **5**, 1, (2011).